FLOWS IN NETWORKS

Flows in Networks

BY

L. R. FORD, Jr.

C–E–I–R, INC.

D. R. FULKERSON

THE RAND CORPORATION

1962

PRINCETON UNIVERSITY PRESS

PRINCETON, NEW JERSEY

Printed in the United States of America

PREFACE

This book presents one approach to that part of linear programming theory that has come to be encompassed by the phrase "transportation problems" or "network flow problems." We use the latter name, not only because it is more nearly suggestive of the mathematical content of the subject, but also because it is less committed to one domain of application. Since many of the applications that are examined have little to do with transportation (and we have not included all the different ways in which the theory has already been used), it seems appropriate not to stress one particular applied area over others.

Just where the study of network flow problems may be said to have originated is a debatable question. Certain static minimal cost transportation models were independently studied by Hitchcock, Kantorovitch, and Koopmans in the 1940's. A few years later, when linear programming began to make itself known as an organized discipline, Dantzig showed how his general algorithm for solving linear programs, the simplex method, could be simplified and made more effective for the special case of transportation models. It would not be inaccurate to say that the subject matter of this book began with the work of these men on the very practical problem of transporting a commodity from certain points of supply to other points of demand in a way to minimize shipping cost. (This problem forms the nucleus of our Chapter III, entitled "Minimal Cost Flow Problems.") However, dismissing the formulational and applied aspects of the subject completely, and with the advantages of hindsight, one can go back a few years earlier to research of König, Egerváry, and Menger on linear graphs, or Hall on systems of distinct representatives for sets, and also relate this work in pure mathematics to the practically oriented subject of flows in networks. We have done this in Chapter II, "Feasibility Theorems and Combinatorial Applications."

One characteristic of the book that has been suggested above should perhaps be made explicit. While this is primarily a book in applied mathematics, we have also included topics that are purely mathematically motivated, together with those that are strictly utilitarian in concept. For this, no apology is intended. We have simply written about mathematics which has interested us, pure or applied.

To carry the historical sketch another (and our last) step back in time might lead one to the Maxwell-Kirchhoff theory of current distribution in

vii

an electrical network. Although this topic is closely related to the subject of the book, we have chosen not to include it. The reason for this is that we have limited the flow problems discussed to purely linear ones and, within this category, to those for which the assumption of integral data in the problem implies the existence of an integral solution. This sub-class of linear flow problems has, we feel, a simple elegance not shared by those outside the class. The first restriction, that of linearity, eliminates the Maxwell-Kirchhoff electrical network problem, which, viewed as a programming problem, becomes one of minimizing a quadratic function subject to linear constraints. The second restriction eliminates, for example, linear problems that involve the simultaneous flow of several commodities, important as these may be in practical applications of linear programming.

There are four chapters in the pages that follow; two of them (Chapters II and III) have been mentioned already. Chapter I, "Static Maximal Flow," studies the problem of maximizing flow from one point to another in a capacity-constrained network. From our point of view, this problem is the most fundamental topic dealt with in the book. Its solution provides a method of attack on the feasibility and combinatorial questions that form the subject of Chapter II, while the simple construction that results, when taken in conjunction with work of Kuhn on the optimal assignment problem, provides the key to the development of the various minimal cost flow methods in Chapter III. In addition, the recent treatment by Gomory and Hu of multi-terminal maximal flows, which is presented in Chapter IV, relies heavily on the central theorem of Chapter I. Thus Chapter I is prerequisite to the others, which are largely independent of each other.

Throughout the book the emphasis is on constructive procedures, even more, on computationally effective ones. Other things being nearly equal, we prefer a constructive proof of a theorem to a non-constructive one, and a constructive proof that leads to an efficient computational scheme is, to our way of thinking, just that much better.

The reader who is familiar with the simplex method of solution for network flow problems will find that this facet of the subject has been omitted in our presentation. For example, the notion of a spanning sub-tree of a network, which would play a fundamental role in the simplex theory, is not introduced until the last chapter, and then for another use. This omission does not reflect an aesthetic judgment on our part; it is, rather, that the more purely combinatorial methods developed here seem to be better computationally and also yield fresh insight into the subject.

ACKNOWLEDGMENTS

It is a pleasure to record our obligation to the following people, with whom we have discussed various parts of the manuscript: S.E.Dreyfus, R.E.Gomory, T.C.Hu, H.J.Ryser, L.S.Shapley, P.Wolfe, and J.W.T. Youngs. We want also to express special appreciation to T.E.Harris and F.S.Ross, who stimulated our interest in the topic of flows in networks, and to G.B.Dantzig, who has provided steady encouragement in our work on this topic over the past several years.

This study was prepared as part of the continuing program of research undertaken for the U.S. Air Force by The RAND Corporation.

CONTENTS

CHAPTER I

STATIC MAXIMAL FLOW

CHAPTER II

FEASIBILITY THEOREMS AND COMBINATORIAL APPLICATIONS

CHAPTER III

MINIMAL COST FLOW PROBLEMS

CHAPTER IV

MULTI-TERMINAL MAXIMAL FLOWS

FLOWS IN NETWORKS

CHAPTER I

STATIC MAXIMAL FLOW

Introduction

The mathematical problem which forms the subject matter of this chapter, that of determining a maximal steady state flow from one point to another in a network subject to capacity limitations on arcs, comes up naturally in the study of transportation or communication networks. It was posed to the authors in the spring of 1955 by T. E. Harris, who, in conjunction with General F. S. Ross (Ret.), had formulated a simplified model of railway traffic flow, and pinpointed this particular problem as the central one suggested by the model [11]. It was not long after this until the main result, Theorem 5.1, which we call the max-flow min-cut theorem, was conjectured and established [4]. A number of proofs of this theorem have since appeared [2, 3, 5, 14]. The constructive proof given in § 5 is the simplest and most revealing of the several known to us.

Sections 1 and 2 discuss networks and flows in networks. There are many alternative ways of formulating the concept of a flow through a network; two of these are described in § 2. After introducing some notation in § 3, and defining the notion of a cut in § 4, we proceed to a statement and proof of the max-flow min-cut theorem in § 5. Sections 6, 7, 9, 10, and 11 amplify and extend this theorem. In § 8, the construction implicit in its proof is detailed and illustrated. This construction, which we call the "labeling process," forms the basis for almost all the algorithms presented later in the book. A consequence of the construction is the integrity theorem (Theorem 8.1), which has been known in connection with similar problems since G. B. Dantzig pointed it out in 1951 [1]. It is this theorem that makes network flow theory applicable in certain combinatorial investigations.

Section 12 provides a brief presentation of duality principles for linear programs. Since no proofs are included, the reader not familiar with linear inequality theory may find this section too brief to be very illuminating. But excellent discussions are available [8, 9, 10]. We include § 12 mainly to note that the max-flow min-cut theorem is a kind of combinatorial counterpart, for the special case of the maximal flow problem, of the more general duality theorem for linear programs.

Section 13 uses the max-flow min-cut theorem to examine maximal flow through a network as a function of a pair of individual arc capacities. The

1

main conclusion here, which may sound empty but is not, is that any two arcs either always reinforce each other or always interfere with each other.

1. Networks

A *directed network* or *directed linear graph* $G = [N; \mathscr{A}]$ consists of a collection N of elements $x, y, \ldots$, together with a subset $\mathscr{A}$ of the ordered pairs (x, y) of elements taken from N. It is assumed throughout that N is a finite set, since our interest lies mainly in the construction of computational procedures. The elements of N are variously called *nodes, vertices, junction points,* or *points*; members of $\mathscr{A}$ are referred to as *arcs, links, branches,* or *edges*. We shall use the node-arc terminology throughout.

A network may be pictured by selecting a point corresponding to each node x of N and directing an arrow from x to y if the ordered pair (x, y) is in $\mathscr{A}$. For example, the network shown in Fig. 1.1 consists of four nodes s, x, y, t, and six arcs (s, x), (s, y), (x, y), (y, x), (x, t) and (y, t).

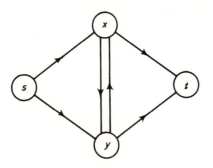

Figure 1.1

Such a network is said to be directed, since each arc carries a specific orientation or direction. Occasionally we shall also consider *undirected networks*, for which the set $\mathscr{A}$ consists of unordered pairs of nodes, or *mixed networks*, in which some arcs are directed, others are not. We can of course picture these in the same way, omitting arrowheads on arcs having no orientation. Until something is said to the contrary, however, each arc of the network will be assumed to have an orientation.

We have not as yet ruled out the possibility of arcs (x, x) leading from a node x to itself, but for our purposes we may as well do so. Thus, all arcs will be supposed to be of the form (x, y) with $x \neq y$. Also, while the existence of at most one arc (x, y) has been postulated, the notion of a network frequently allows multiple arcs joining x to y. Again, for most of the problems we shall consider, this added generality gains nothing, and so we shall continue to think of at most one arc leading from any node to another, unless an explicit statement is made to the contrary.

§1. NETWORKS

Let $x_1, x_2, \ldots, x_n$ $(n \geqslant 2)$ be a sequence of distinct nodes of a network such that (x_i, x_{i+1}) is an arc, for each $i = 1, \ldots, n - 1$. Then the sequence of nodes and arcs

(1.1)
$$x_1, (x_1, x_2), x_2, \ldots, (x_{n-1}, x_n), x_n$$

is called a *chain*; it leads from x_1 to x_n. Sometimes, for emphasis, we call (1.1) a *directed chain*. If the definition of a chain is altered by stipulating that $x_n = x_1$, then the displayed sequence is a *directed cycle*. For example, in the network of Fig. 1.1, the chain $s, (s, x), x, (x, t), t$ leads from s to t; this network contains just one directed cycle, namely, $x, (x, y), y, (y, x), x$.

Let $x_1, x_2, \ldots, x_n$ be a sequence of distinct nodes having the property that either (x_i, x_{i+1}) or (x_{i+1}, x_i) is an arc, for each $i = 1, \ldots, n - 1$. Singling out, for each i, one of these two possibilities, we call the resulting sequence of nodes and arcs a *path from x_1 to x_n*. Thus a path differs from a chain by allowing the possibility of traversing an arc in a direction opposite to its orientation in going from x_1 to x_n. (For undirected networks, the two notions coincide.) Arcs (x_i, x_{i+1}) that belong to the path are *forward* arcs of the path; the others are *reverse* arcs. For example, the sequence $s, (s, y), y, (x, y), x, (x, t), t$ is a path from s to t in Fig. 1.1; it contains the forward arcs $(s, y), (x, t)$ and the reverse arc (x, y). If, in the definition of path, we stipulate that $x_1 = x_n$, then the resulting sequence of nodes and arcs is a *cycle*.

We may shorten the notation and refer unambiguously to the chain $x_1, x_2, \ldots, x_n$. Occasionally we shall also refer to the path $x_1, x_2, \ldots, x_n$; then it is to be understood that some selection of arcs has tacitly been made.

Given a network $[N; \mathscr{A}]$, one can form a *node-arc incidence matrix* as follows. List the nodes of the network vertically, say, the arcs horizontally, and record, in the column corresponding to arc (x, y), a 1 in the row corresponding to node x, a -1 in the row corresponding to y, and zeros elsewhere. For example, the network of Fig. 1.1 has incidence matrix

$$
\begin{array}{c}
\\ s \\ x \\ y \\ t
\end{array}
\begin{array}{cccccc}
(s, x) & (s, y) & (x, y) & (y, x) & (x, t) & (y, t) \\
\left[\begin{array}{cccccc}
1 & 1 & 0 & 0 & 0 & 0 \\
-1 & 0 & 1 & -1 & 1 & 0 \\
0 & -1 & -1 & 1 & 0 & 1 \\
0 & 0 & 0 & 0 & -1 & -1
\end{array}\right].
\end{array}
$$

Clearly, all information about the structure of a network is embodied in its node-arc incidence matrix.

If $x \in N$, we let $A(x)$ ("after x") denote the set of all $y \in N$ such that $(x, y) \in \mathscr{A}$:

(1.2)
$$A(x) = \{y \in N \,|\, (x, y) \in \mathscr{A}\}.$$

Similarly, we let $B(x)$ ("before x") denote the set of all $y \in N$ such that $(y, x) \in \mathscr{A}$:

(1.3) $$B(x) = \{y \in N | (y, x) \in \mathscr{A}\}.$$

For example, in the network of Fig. 1.1,

$$A(s) = \{x, y\}, \qquad B(s) = \varnothing \text{ (the empty set)}.$$

We shall on occasion require some other notions concerning networks. These will be introduced as needed.

2. Flows in networks

Given a network $G = [N; \mathscr{A}]$, suppose that each arc $(x, y) \in \mathscr{A}$ has associated with it a non-negative real number $c(x, y)$. We call $c(x, y)$ the *capacity* of the arc (x, y); it may be thought of intuitively as representing the maximal amount of some commodity that can arrive at y from x per unit time. The function c from $\mathscr{A}$ to non-negative reals is the *capacity function*. (Sometimes it will be convenient to allow infinite arc capacities also.)

The fundamental notion underlying most of the topics treated subsequently is that of a static or steady state flow through a network, which we now proceed to formulate. (Since dynamic flows will not be discussed until Chapter III, the qualifying phrase "static" or "steady state" will usually be omitted.)

Let s and t be two distinguished nodes of N. A *static flow of value v from s to t* in $[N; \mathscr{A}]$ is a function f from $\mathscr{A}$ to non-negative reals that satisfies the linear equations and inequalities

(2.1) $$\sum_{y \in A(x)} f(x, y) - \sum_{y \in B(x)} f(y, x) = \begin{cases} v, & x = s, \\ 0, & x \neq s, t, \\ -v, & x = t, \end{cases}$$

(2.2) $$f(x, y) \leqslant c(x, y) \qquad \text{all } (x, y) \in \mathscr{A}.$$

We call s the *source*, t the *sink*, and other nodes *intermediate*. Thus if the *net flow out of x* is defined to be

$$\sum_{y \in A(x)} f(x, y) - \sum_{y \in B(x)} f(y, x),$$

then the equations (2.1) may be verbalized by saying that the net flow out of the source is v, the net flow out of the sink is $-v$ (or the net flow into the sink is v), whereas the net flow out of an intermediate node is zero. An equation of the latter kind will be called a *conservation equation*.

When necessary to avoid ambiguity, we shall denote the value of a flow f by $v(f)$. Notice that a flow f from s to t of value v is a flow from t to s of value $-v$.

An example of a flow from s to t is shown in Fig. 2.1, where it is assumed that arc capacities are sufficiently large so that none are violated. The value of this flow is 3.

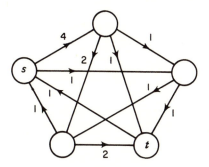

Figure 2.1

Given a flow f, we refer to $f(x, y)$ as the *arc flow* $f(x, y)$ or the *flow in arc* (x, y). Each arc flow $f(x, y)$ occurs in precisely two equations of (2.1), and has a coefficient 1 in the equation corresponding to node x, a coefficient -1 in the equation corresponding to node y. In other words, the coefficient matrix of equations (2.1), apart from the column corresponding to v, is the node-arc incidence matrix of the network. (By adding the special arc (t, s) to the network, allowing multiple arcs if necessary, a non-negative flow value v can be thought of as the "return flow" in (t, s), and all equations taken as conservation equations.)

A few observations. There is no question concerning the existence of flows, since $f = 0$, $v = 0$ satisfy (2.1) and (2.2). Also, while we have assumed that $\mathscr{A}$ may be a subset of the ordered pairs (x, y), $x \neq y$, with the capacity function c non-negative on $\mathscr{A}$, we could extend $\mathscr{A}$ to all ordered pairs by taking $c = 0$ outside of $\mathscr{A}$, or we could assume strict positivity of c by deleting from $\mathscr{A}$ arcs having zero capacity. Finally, the set of equations (2.1) is redundant, since adding the rows of its coefficient matrix produces the zero vector. Thus we could omit any one of the equations without loss of generality. We prefer, however, to retain the one-one correspondence between equations and nodes.

The static maximal flow problem is that of maximizing the variable v subject to the flow constraints (2.1) and (2.2). Before proceeding to this problem, it is worth while to point out an alternative formulation that is informative and will be useful in later contexts. This might be termed the arc-chain notion of a flow from s to t.

Suppose that $A_1, \ldots, A_m$ is an enumeration of the arcs of a network, the arc A_i having capacity $c(A_i)$; and let $C_1, \ldots, C_n$ be a list of all directed

5

chains from s to t. Form the m by n incidence matrix (a_{ij}) of arcs versus chains by defining

$$(2.3) \qquad a_{ij} = \begin{cases} 1, & \text{if } A_i \in C_j, \\ 0, & \text{otherwise.} \end{cases}$$

Now let h be a function from the set of chains $C_1, \ldots, C_n$ to non-negative reals that satisfies the inequalities

$$(2.4) \qquad \sum_{j=1}^{n} a_{ij} h(C_j) \leqslant c(A_i), \qquad i = 1, \ldots, m.$$

We refer to h as a *flow from s to t in arc-chain form*, and call $h(C_j)$ a *chain flow* or the *flow in chain C_j*. The *value* of h is

$$(2.5) \qquad v(h) = \sum_{j=1}^{n} h(C_j).$$

When we need to distinguish the two notions of a flow from s to t thus far introduced, we shall call a function f from the set of arcs to non-negative reals which satisfies (2.1) and (2.2) for some v, a *flow from s to t in node-arc form*. There will usually be no need for the distinction, since we shall work almost exclusively with node-arc flows after this section.

Let us explore the relationship between these two formulations of the intuitive notion of a flow. Suppose that $x_1, \ldots, x_l$ is a list of the nodes, and let (b_{ki}), $k = 1, \ldots, l$, $i = 1, \ldots, m$, be the node-arc incidence matrix introduced earlier. Thus

$$(2.6) \qquad b_{ki} = \begin{cases} 1, & \text{if } A_i = (x_k, y), \\ -1, & \text{if } A_i = (y, x_k), \\ 0, & \text{otherwise.} \end{cases}$$

Then

$$b_{ki} a_{ij} = \begin{cases} 1, & \text{if } A_i = (x_k, y) \text{ and } A_i \in C_j, \\ -1, & \text{if } A_i = (y, x_k) \text{ and } A_i \in C_j, \\ 0, & \text{otherwise,} \end{cases}$$

and it follows that

$$(2.7) \qquad \sum_{i=1}^{m} b_{ki} a_{ij} = \begin{cases} 1, & \text{if } x_k = s, \\ -1, & \text{if } x_k = t, \\ 0, & \text{otherwise.} \end{cases}$$

If h is a flow from s to t in arc-chain form, and if we define

$$(2.8) \qquad f(A_i) = \sum_{j=1}^{n} a_{ij} h(C_j), \qquad i = 1, \ldots, m,$$

then f is a flow from s to t in node-arc form, and $v(f) = v(h)$. For, by (2.4) and (2.8),

$$f(A_i) \leqslant c(A_i),$$

6

and by (2.7),

$$\sum_{i=1}^{m} b_{ki} f(A_i) = \sum_{i=1}^{m} \sum_{j=1}^{n} b_{ki} a_{ij} h(C_j)$$

$$= \sum_{j=1}^{n} \left(\sum_{i=1}^{m} b_{ki} a_{ij} \right) h(C_j)$$

$$= \begin{cases} \sum_{j=1}^{n} h(C_j), & \text{if } x_k = s, \\ -\sum_{j=1}^{n} h(C_j), & \text{if } x_k = t, \\ 0, & \text{otherwise.} \end{cases}$$

But these are precisely equations (2.1) for the function f and $v = \sum_{j=1}^{n} h(C_j)$.

On the other hand, we can start with a flow f in node-arc form having value v, and obtain from it a flow h in arc-chain form having value $v(h) \geqslant v$. Intuitively, the reason the inequality now appears is that the node-arc formulation permits flow along chains from t to s.

There are various ways of obtaining such an arc-chain flow h from a given node-arc flow f. One way is as follows. Define

$$(2.9) \qquad h(C_j) = \min_{A_i \in C_j} f_j(A_i), \qquad j = 1, \ldots, n,$$

where

$$(2.10) \qquad f_j(A_i) = f(A_i) - \sum_{p=1}^{j-1} a_{ip} h(C_p), \qquad j = 1, \ldots, n+1.$$

In words, look at the first chain C_1, reduce $f_1 = f$ by as much as possible (retaining non-negativity of arc flows) on arcs of C_1; this yields f_2. The process is then repeated with C_2 and f_2, and so on until all chains have been examined. It follows that f_{j+1} is a node-arc flow from s to t having value $v(f_{j+1}) = v - \sum_{p=1}^{j} h(C_p)$, since

$$\sum_{i=1}^{m} b_{ki} f_{j+1}(A_i) = \sum_{i=1}^{m} b_{ki} f(A_i) - \sum_{i=1}^{m} \sum_{p=1}^{j} b_{ki} a_{ip} h(C_p),$$

$$= \begin{cases} v - \sum_{p=1}^{j} h(C_p), & \text{if } x_k = s, \\ -v + \sum_{p=1}^{j} h(C_p), & \text{if } x_k = t, \\ 0, & \text{otherwise.} \end{cases}$$

7

Moreover, $f_{j+1}(A_i) \leqslant f_j(A_i)$, all A_i, and $f_{j+1}(A_i) = 0$ for some $A_i \in C_j$. Hence the node-arc flow f_{n+1} vanishes on some arc of every chain from s to t. This implies that $v(f_{n+1}) \leqslant 0$, as the following lemma shows.

LEMMA 2.1.   *If f is a node-arc flow from s to t having value $v(f) > 0$, then there is a chain from s to t such that $f > 0$ on all arcs of this chain.*

PROOF.   Let X be the set of nodes defined recursively by the rules
(a) $s \in X$,
(b) if $x \in X$, and if $f(x, y) > 0$, then $y \in X$.
We assert that $t \in X$. For, suppose not. Then, summing the equations (2.1) over $x \in X$, and noting cancellations, we have

$$v(f) = \sum_{\substack{x \in X \\ y \notin X}} [f(x, y) - f(y, x)].$$

But by (b), if (x, y) is an arc with $x \in X$, $y \notin X$, then $f(x, y) = 0$. This and the last displayed equation contradict $v(f) > 0$. Thus $t \in X$. But for any $x \in X$, the definition of X shows that there is a chain from s to x such that $f > 0$ on arcs of this chain. Hence there is a chain from s to t with this property.

It follows from the lemma that the value of f_{n+1} is non-positive, that is

$$v(f_{n+1}) = v - \sum_{p=1}^{n} h(C_p) \leqslant 0.$$

Consequently $v(h) \geqslant v$. This proves

THEOREM 2.2.   *If h is an arc-chain flow from s to t, then f defined by (2.8) is a node-arc flow from s to t and $v(f) = v(h)$. On the other hand, if f is a node-arc flow from s to t, then h defined by (2.9) and (2.10) is an arc-chain flow from s to t, and $v(h) \geqslant v(f)$.*

A consequence of Theorem 2.2 is that it is immaterial whether the maximal flow problem is formulated in terms of the node-arc incidence matrix or the arc-chain incidence matrix. Thus, for example, since arcs of the form (x, s) or (t, x) can be deleted from $\mathscr{A}$ without changing the list of chains from s to t, we may always suppose in either formulation of the maximal flow problem that all source arcs point out from the source, and all sink arcs point into the sink. (For such networks, one has $v(h) = v(f)$ in the second part of Theorem 2.2 as well as the first part.)

A function h defined from f as in (2.9) and (2.10) will be termed a *chain decomposition* of f. A chain decomposition of f will, in general, depend on the ordering of the chains. For example, if in Fig. 1.1 we take $f = 1$ on all arcs, and take $C_1 = (s, x, t)$, $C_2 = (s, y, t)$, $C_3 = (s, x, y, t)$, $C_4 = (s, y, x, t)$, then $h(C_1) = h(C_2) = 1$, $h(C_3) = h(C_4) = 0$. But, examining the chains in reverse order would lead to $h(C_4) = h(C_3) = 1$, $h(C_2) = h(C_1) = 0$.

8

From the computational point of view, one would certainly suppose the node-arc formulation of the maximal flow problem to be preferable for most networks, since the number of chains from s to t is likely to be large compared to the number of nodes or the number of arcs. A computing procedure that required as a first step the enumeration of all chains from s to t would be of little value. There are less obvious reasons why the node-arc formulation is to be preferred from the theoretical point of view as well.†

3. Notation

To simplify the notation, we adopt the following conventions. If X and Y are subsets of N, let (X, Y) denote the set of all arcs that lead from $x \in X$ to $y \in Y$; and, for any function g from $\mathscr{A}$ to reals, let

$$(3.1) \qquad \sum_{(x,y) \in (X,Y)} g(x, y) = g(X, Y).$$

Similarly, when dealing with a function h defined on the nodes of N, we put

$$(3.2) \qquad \sum_{x \in X} h(x) = h(X).$$

We customarily denote a set consisting of one element by its single element. Thus if X contains the single node x, we write (x, Y), $g(x, Y)$, and so on.

Set unions, intersections, and differences will be denoted by $\cup$, $\cap$, and $-$, respectively. Thus $X \cup Y$ is the set of nodes in X or in Y, $X \cap Y$ the set of nodes in both X and Y, and $X - Y$ the set of nodes in X but not in Y. We use $\subseteq$ for set inclusion, and $\subset$ for proper inclusion. Complements of sets will be denoted by barring the appropriate symbol. For instance, the complement of X in N is $\overline{X} = N - X$.

Thus, if $X, Y, Z \subseteq N$, then

$$(3.3) \qquad g(X, Y \cup Z) = g(X, Y) + g(X, Z) - g(X, Y \cap Z),$$

$$(3.4) \qquad g(Y \cup Z, X) = g(Y, X) + g(Z, X) - g(Y \cap Z, X).$$

Hence if Y and Z are disjoint,

$$g(X, Y \cup Z) = g(X, Y) + g(X, Z),$$

$$g(Y \cup Z, X) = g(Y, X) + g(Z, X).$$

† Two comments are in order here. First, one can describe a computing procedure for the arc-chain formulation of the maximal flow problem that does not require an explicit enumeration of all chains [6]. Second, a strong theoretical reason for adopting the node-arc formulation, nonetheless, is that the node-arc incidence matrix has a desirable property not shared by the arc-chain incidence matrix. This is the unimodularity property, that is, every submatrix has determinant ± 1 or 0. See [12] for a full discussion of this property and its implications for linear programming problems.

Notice that

$$(B(x), x) = (N, x),$$

$$(x, A(x)) = (x, N),$$

and

$$g(N, X) = \sum_{x \in X} g(N, x) = \sum_{x \in X} g(B(x), x),$$

$$g(X, N) = \sum_{x \in X} g(x, N) = \sum_{x \in X} g(x, A(x)).$$

Later on (Chapter II) we shall use the notation $|X|$ to denote the number of elements in an arbitrary set X.

4. Cuts

Progress toward a solution of the maximal network flow problem is made with the recognition of the importance of certain subsets of arcs, which we shall call cuts. A *cut* $\mathscr{C}$ in $[N; \mathscr{A}]$ *separating s and t* is a set of arcs $(X, \overline{X})$ where $s \in X$, $t \in \overline{X}$. The *capacity* of the cut $(X, \overline{X})$ is $c(X, \overline{X})$.

For example, the set of arcs $\mathscr{C} = \{(s, y), (x, y), (x, t)\}$ with $X = \{s, x\}$, is a cut in the network of Fig. 1.1 separating s and t.

Notice that any chain from s to t must contain some arc of every cut $(X, \overline{X})$. For let $x_1, x_2, \ldots, x_n$ be a chain with $x_1 = s$, $x_n = t$. Since $x_1 \in X$, $x_n \in \overline{X}$, there is an x_i $(1 \leqslant i < n)$ with $x_i \in X$, $x_{i+1} \in \overline{X}$. Hence the arc (x_i, x_{i+1}) is a member of the cut $(X, \overline{X})$. It follows that if all arcs of a cut were deleted from the network, there would be no chain from s to t and the maximal flow value for the new network would be zero.

Since a cut blocks all chains from s to t, it is intuitively clear (and indeed obvious in the arc-chain version of the problem) that the value v of a flow f cannot exceed the capacity of any cut, a fact that we now prove from (2.1) and (2.2).

LEMMA 4.1. *Let f be a flow from s to t in a network $[N; \mathscr{A}]$, and let f have value v. If $(X, \overline{X})$ is a cut separating s and t, then*

$$(4.1) \qquad v = f(X, \overline{X}) - f(\overline{X}, X) \leqslant c(X, \overline{X}).$$

PROOF. The equality of (4.1) was actually proved in Lemma 2.1. We re-prove it here, using the notation introduced in the preceding section. Since f is a flow, f satisfies the equations

$$f(s, N) - f(N, s) = v,$$
$$f(x, N) - f(N, x) = 0, \qquad\qquad x \neq s, t,$$
$$f(t, N) - f(N, t) = -v.$$

Now sum these equations over $x \in X$. Since $s \in X$ and $t \in \overline{X}$, the result is

$$v = \sum_{x \in X} (f(x, N) - f(N, x)) = f(X, N) - f(N, X).$$

Writing $N = X \cup \overline{X}$ in this equality yields

$$v = f(X, X \cup \overline{X}) - f(X \cup \overline{X}, X)$$
$$= f(X, X) + f(X, \overline{X}) - f(X, X) - f(\overline{X}, X),$$

thus verifying the equality in (4.1). Since $f(\overline{X}, X) \geqslant 0$ and $f(X, \overline{X}) \leqslant c(X, \overline{X})$ by virtue of (2.2), the inequality of (4.1) follows immediately.

In words, the equality of (4.1) states that the value of a flow from s to t is equal to the net flow across any cut separating s and t.

5. Maximal flow

We are now in a position to state and prove the fundamental result concerning maximal network flow [4, 5].

THEOREM 5.1. (Max-flow min-cut theorem.) *For any network the maximal flow value from s to t is equal to the minimal cut capacity of all cuts separating s and t.*

Before proving Theorem 5.1, we illustrate it with an example. Consider the network of Fig. 1.1 with capacity function c and flow f as indicated in Fig. 5.1, $c(x, y)$ being the first member of the pair of numbers written

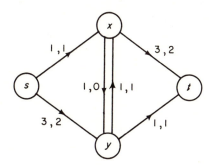

Figure 5.1

adjacent to arc (x, y), and $f(x, y)$ the second. Here the flow value is 3. Since the cut composed of arcs (s, x), (y, x), and (y, t) also has capacity 3, it follows from Lemma 4.1 that the flow is maximal and the cut minimal.

PROOF OF THEOREM 5.1. By Lemma 4.1, it suffices to establish the existence of a flow f and a cut $(X, \overline{X})$ for which equality of flow value and cut capacity holds. We do this by taking a maximal flow f (clearly such exists) and defining, in terms of f, a cut $(X, \overline{X})$ such that

$$f(X, \overline{X}) = c(X, \overline{X}),$$
$$f(\overline{X}, X) = 0,$$

so that equality holds throughout (4.1).

11

Thus, let f be a maximal flow. Using f, define the set X recursively as follows:

(a) $s \in X$;
(b) if $x \in X$ and $f(x, y) < c(x, y)$, then $y \in X$;
    if $x \in X$ and $f(y, x) > 0$, then $y \in X$.

We assert that $t \in \bar{X}$. For, suppose not. It then follows from the definition of X that there is a path from s to t, say

$$s = x_1, x_2, \ldots, x_n = t,$$

having the property that for all forward arcs (x_i, x_{i+1}) of the path,

$$f(x_i, x_{i+1}) < c(x_i, x_{i+1}),$$

whereas for all reverse arcs (x_{i+1}, x_i) of the path,

$$f(x_{i+1}, x_i) > 0.$$

Let ε_1 be the minimum of $c - f$ taken over all forward arcs of the path, ε_2 the minimum of f taken over all reverse arcs, and let $\varepsilon = \min(\varepsilon_1, \varepsilon_2) > 0$. Now alter the flow f as follows: increase f by ε on all forward arcs of the path, and decrease f by ε on all reverse arcs. It is easily checked that the new function thus defined is a flow from s to t having value $v + \varepsilon$. But then f is not maximal, contrary to our assumption, and thus $t \in \bar{X}$.

Consequently $(X, \bar{X})$ is a cut separating s and t. Moreover, from the definition of X, it follows that

$$f(x, \bar{x}) = c(x, \bar{x}) \qquad \text{for } (x, \bar{x}) \in (X, \bar{X}),$$
$$f(\bar{x}, x) = 0 \qquad \text{for } (\bar{x}, x) \in (\bar{X}, X),$$

since otherwise $\bar{x}$ would be in X. Thus

$$f(X, \bar{X}) = c(X, \bar{X}), \qquad f(\bar{X}, X) = 0,$$

so that equality holds in 4.1.

Several corollaries can be gleaned from Lemma 4.1, Theorem 5.1, and its proof.

We shall call a path from s to t a *flow augmenting path* with respect to a flow f provided that $f < c$ on forward arcs of the path, and $f > 0$ on reverse arcs of the path. Then we have

COROLLARY 5.2. *A flow f is maximal if and only if there is no flow augmenting path with respect to f.*

PROOF. If f is maximal, then clearly no flow augmenting path exists. Suppose, conversely, that no flow augmenting path exists. Then the set X defined recursively using f as in the proof of Theorem 5.1 cannot contain the sink t. Hence, as in the proof of Theorem 5.1, $(X, \bar{X})$ is a cut separating s and t having capacity equal to the value of f. Consequently f is maximal.

Corollary 5.2 is of fundamental importance in the study of network flows. It says, in essence, that in order to increase the value of a flow, it suffices to look for improvements of a very restricted kind.

We say that an arc (x, y) is *saturated* with respect to a flow f if $f(x, y) = c(x, y)$ and is *flowless* with respect to f if $f(x, y) = 0$. Thus an arc that is both saturated and flowless has zero capacity. Corollary 5.3 characterizes a minimal cut in terms of these notions.

COROLLARY 5.3. *A cut $(X, \overline{X})$ is minimal if and only if every maximal flow f saturates all arcs of $(X, \overline{X})$ whereas all arcs of $(\overline{X}, X)$ are flowless with respect to f.*

Using Corollary 5.3 it is easy to prove

COROLLARY 5.4. *Let $(X, \overline{X})$ and $(Y, \overline{Y})$ be minimal cuts. Then $(X \cup Y, \overline{X \cup Y})$ and $(X \cap Y, \overline{X \cap Y})$ are also minimal cuts.*

The following theorem shows that the minimal cut $(X, \overline{X})$ singled out in the proof of Theorem 5.1 does not, in actuality, depend on the maximal flow f.

THEOREM 5.5. *Let $(Y, \overline{Y})$ be any minimal cut, let f be a maximal flow, and let $(X, \overline{X})$ be the minimal cut defined relative to f in the proof of Theorem 5.1. Then $X \subseteq Y$.*

PROOF. Suppose that X is not included in Y. Then $X \cap Y \subset X$, and $(X \cap Y, \overline{X \cap Y})$ is a minimal cut by Corollary 5.4. Let x be a node in X that is not in $X \cap Y$. Since $x \in X$ and $x \neq s$, there is a path from s to x, say $s = x_1, x_2, \ldots, x_k = x$, such that each forward arc of the path is unsaturated with respect to f, while each reverse arc carries positive flow. But since $s \in X \cap Y$ and $x \in \overline{X \cap Y}$, there is a pair x_i, x_{i+1} $(1 \leqslant i < k)$ such that $x_i \in X \cap Y$, $x_{i+1} \in \overline{X \cap Y}$. If (x_i, x_{i+1}) is a forward arc of the path, then $f(x_i, x_{i+1}) < c(x_i, x_{i+1})$, contradicting Corollary 5.3. Similarly if (x_{i+1}, x_i) is a reverse arc of the path, Corollary 5.3 is contradicted. Hence $X \subseteq Y$.

Thus if $(X_i, \overline{X}_i)$, $i = 1, \ldots, m$, are all the minimal cuts separating source and sink, the set X defined relative to a particular maximal flow in the proof of Theorem 5.1 is the intersection of all X_i and hence does not depend on the selection of the flow.

Although the minimal cut $(X, \overline{X})$ was picked out in the proof of Theorem 5.1 by a recursive definition of the source set X, symmetrically we could have generated a minimal cut $(Y, \overline{Y})$ by defining its sink set $\overline{Y}$ in terms of a maximal flow f as follows:

(a') $t \in \overline{Y}$;
(b') if $y \in \overline{Y}$ and $f(x, y) < c(x, y)$, then $x \in \overline{Y}$;
   if $y \in \overline{Y}$ and $f(y, x) > 0$, then $x \in \overline{Y}$.

Equivalently, one can think of reversing all arc orientations and arc flows, interchanging source and sink so that t becomes the source, s the sink, and then use the definition given in the proof of Theorem 5.1 to construct $\bar{Y}$. Again, although its definition is made relative to a particular maximal flow, the set $\bar{Y}$ does not actually depend on the selection, since $\bar{Y}$ is the intersection of the sink sets $\bar{X}_i$ of all minimal cuts $(X_i, \bar{X}_i)$.

Using both definitions, we can state a criterion for uniqueness of a minimal cut.

THEOREM 5.6.   *Let X be the set of nodes defined in the proof of Theorem 5.1, let $\bar{Y}$ be the set defined above, and assume that c is strictly positive. The minimal cut $(X, \bar{X})$ is unique if and only if $(X, \bar{X}) = (Y, \bar{Y})$.*

PROOF.   We must show that if $(X, \bar{X}) = (Y, \bar{Y})$, and if $(Z, \bar{Z})$ is any minimal cut, then $(X, \bar{X}) = (Z, \bar{Z})$.

First note that if $(X, \bar{X}) = (Y, \bar{Y})$, then both equal $(X, \bar{Y})$. For, $X \subseteq Y$ by Theorem 5.5, hence $(X, \bar{Y}) \subseteq (Y, \bar{Y})$. On the other hand, if $(u, v) \in (X, \bar{X}) = (Y, \bar{Y})$, then $u \in X$ and $v \in \bar{Y}$, so $(u, v) \in (X, \bar{Y})$.

For the arbitrary minimal cut $(Z, \bar{Z})$, we have, again by Theorem 5.5 and its analogue for $(Y, \bar{Y})$, that $X \subseteq Z$, $\bar{Y} \subseteq \bar{Z}$. Thus $(X, \bar{Y}) \subseteq (Z, \bar{Y}) \subseteq (Z, \bar{Z})$. Hence $c(X, \bar{Y}) \leqslant c(Z, \bar{Z})$. Now if $(X, \bar{Y}) \subset (Z, \bar{Z})$, then either some arcs of $(Z, \bar{Z})$ have zero capacity, contradicting our assumption $c > 0$, or $c(X, \bar{Y}) < c(Z, \bar{Z})$, contradicting the minimality of $(Z, \bar{Z})$. Thus $(X, \bar{X}) = (X, \bar{Y}) = (Z, \bar{Z})$.

Notice that Theorem 5.6 is not valid if the assumption $c > 0$ is relaxed to $c \geqslant 0$. For instance, in the network shown in Fig. 5.2, $X = \{s\}$, $\bar{Y} = \{t\}$,

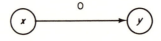

Figure 5.2

and $(X, \bar{X}) = (Y, \bar{Y}) = (s, t)$. However, $(Z, \bar{Z})$ with $Z = \{s, x\}$ is another minimal cut that contains both arcs.

6. Disconnecting sets and cuts

We have characterized cuts as sets of arcs of the form $(X, \bar{X})$ with $s \in X$, $t \in \bar{X}$, and have noted that a cut blocks all chains from s to t. Thus if we call a set of arcs a *disconnecting set* if it has the chain blocking property, then a cut is a disconnecting set. The converse, however, is not necessarily true. For example, the set of all arcs in a network is a disconnecting set, but may not be a cut.

That every disconnecting set contains a cut can be seen easily as follows. Let $\mathscr{D}$ denote the disconnecting set, and define a subset X of nodes by the rule

(a) $s \in X$;

(b) if $x \in X$ and $(x, y) \in \mathscr{A} - \mathscr{D}$, then $y \in X$.

It is clear that $t \in \overline{X}$ and $(X, \overline{X}) \subseteq \mathscr{D}$. Notice that if $\mathscr{D}$ is a proper disconnecting set, that is, a disconnecting set whose proper subsets are not disconnecting, then $(X, \overline{X}) = \mathscr{D}$. Thus every proper disconnecting set is a cut. The converse may not hold, though. For example, in Fig. 5.2, the cut $(X, \overline{X})$ with $X = \{s, x\}$ is not a proper disconnecting set.

We may summarize the discussion thus far by saying:

(1) the class of proper disconnecting sets is included in the class of cuts, which, in turn, is included in the class of disconnecting sets, and that each of these inclusions may be proper;

(2) every disconnecting set contains a cut.

It follows that the notion of a cut could be replaced by either that of disconnecting set or proper disconnecting set in the statement of the max-flow min-cut theorem.

We have chosen to focus attention on cuts rather than disconnecting sets because the former are more convenient to work with when dealing with flows in node-arc form; the latter are convenient for an arc-chain formulation of the maximal flow problem. (See [4], where a proof of Theorem 5.1 which uses the arc-chain formulation is given.)

Notice that, in any case, restricting attention to proper disconnecting sets is as far as one can go in narrowing the class of sets of arcs that require consideration, since every proper disconnecting set of a network has minimal capacity for some capacity function: for instance, $c(x, y) = 1$ if $(x, y) \in \mathscr{D}$, $c(x, y) = \infty$ otherwise, singles out the proper disconnecting set $\mathscr{D}$ as the unique minimal cut.

7. Multiple sources and sinks

Although the assumption has been that the network has a single source and single sink, it is easy to see that the situation in which there are multiple sources and sinks, with flow permitted from any source to any sink, presents nothing new, since the adjunction of two new nodes and several arcs to the multiple source, multiple sink network reduces the problem to the case of a single source and sink.

15

In more detail, suppose that the nodes N of a network $[N; \mathscr{A}]$ are partitioned into three sets:

S (the set of sources),
T (the set of sinks),
R (the set of intermediate nodes),

and consider the problem of finding a maximal flow from S to T.

A flow from S to T may be thought of as a real valued function f defined on $\mathscr{A}$ that satisfies

$$(7.1) \qquad f(x, N) - f(N, x) = 0 \qquad \text{for } x \in R,$$

$$(7.2) \qquad 0 \leqslant f(x, y) \leqslant c(x, y) \qquad \text{for } (x, y) \in \mathscr{A},$$

the flow value being

$$(7.3) \qquad v = f(S, N) - f(N, S).$$

Extend $[N; \mathscr{A}]$ to a network $[N^*; \mathscr{A}^*]$ by adjoining two nodes u, v and all arcs (u, S), (T, v), and extend the capacity function c defined on $\mathscr{A}$ to c^* defined on $\mathscr{A}^*$ by

$$c^*(u, x) = \infty, \qquad\qquad x \in S,$$
$$c^*(x, v) = \infty, \qquad\qquad x \in T,$$
$$c^*(x, y) = c(x, y), \qquad\qquad (x, y) \in \mathscr{A}.$$

Thus the restriction f of a flow f^* from u to v in $[N^*; \mathscr{A}^*]$ is a flow from S to T in $[N; \mathscr{A}]$. Vice versa, a flow f from S to T in $[N; \mathscr{A}]$ can be extended to a flow f^* from u to v in $[N^*; \mathscr{A}^*]$ by defining

$$f^*(u, x) = f(x, N) - f(N, x), \qquad\qquad x \in S,$$
$$f^*(x, v) = f(N, x) - f(x, N), \qquad\qquad x \in T,$$
$$f^*(x, y) = f(x, y), \qquad\qquad \text{otherwise.}$$

Consequently the maximal flow problem from S to T in $[N; \mathscr{A}]$ is equivalent to a single source, single sink problem in the extended network.

Relevant cuts for the case of many sources S and sinks T are those separating S and T: that is, a set of arcs $(X, \overline{X})$ with $S \subseteq X$, $T \subseteq \overline{X}$. Or, in terms of disconnecting sets, the appropriate notion would be a set of arcs that blocks all chains from S to T. The max-flow min-cut theorem and its corollaries, as well as the other theorems of § 5, remain valid, *mutatis mutandis*, as can be seen either from the equivalent extended problem or by making slight changes in the proofs throughout.

The situation in which there are several sources and sinks, but in which certain sources can "ship" only to certain sinks, is distinctly different. For such a problem, which might be thought of in terms of the simultaneous flow of several commodities, the maximal flow value can be less than the minimal disconnecting set capacity. Here a disconnecting set means a collection of arcs that blocks all chains from sources to corresponding sinks. For example, consider the network shown in Fig. 7.1 with sources

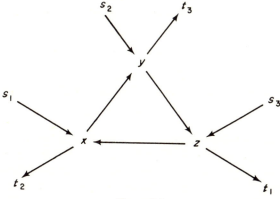

Figure 7.1

s_1, s_2, s_3, and sinks t_1, t_2, t_3. Each arc has unit capacity. Assume that s_i, t_i $(i = 1, 2, 3)$ are the source and sink for commodity i. Then the maximal flow value is $3/2$, obtained by sending a half unit of commodity i along the unique chain from s_i to t_i. However, the arcs (x, y) and (y, z) are a minimal disconnecting set having capacity 2.

8. The labeling method for solving maximal flow problems

Under mild restrictions on the capacity function, the proof of the max-flow min-cut theorem given in § 5 provides a simple and efficient algorithm for constructing a maximal flow and minimal cut in a network [5].

The algorithm may be started with the zero flow. The computation then progresses by a sequence of "labelings" (*Routine* A below), each of which either results in a flow of higher value (*Routine* B below) or terminates with the conclusion that the present flow is maximal.

To ensure termination, it will be assumed that the capacity function c is integral valued. This is not an important restriction computationally, since a problem with rational arc capacities can be reduced to the case of integral capacities by clearing fractions, and of course, for computational purposes, confining attention to rational numbers is really no restriction.

Given an integral flow f, we proceed to assign labels to nodes of the network, a label having one of the forms (x^+, ε) or (x^-, ε), where $x \in N$ and ε is a positive integer or ∞, according to the rules delineated in *Routine* A.

During *Routine* A, a node is considered to be in one of three states: unlabeled, labeled and scanned, or labeled and unscanned. Initially all nodes are unlabeled.

Routine A (labeling process). First the source s receives the label $(-, \varepsilon(s) = \infty)$. (The source is now labeled and unscanned; all other nodes are unlabeled.) In general, select any labeled, unscanned node x. Suppose it is labeled $(z^{\pm}, \varepsilon(x))$. To all nodes y that are unlabeled, and such that $f(x, y) < c(x, y)$, assign the label $(x^{+}, \varepsilon(y))$, where

(8.1) $$\varepsilon(y) = \min\left[\varepsilon(x), c(x, y) - f(x, y)\right].$$

(Such y are now labeled and unscanned.) To all nodes y that are now unlabeled, and such that $f(y, x) > 0$, assign the label $(x^{-}, \varepsilon(y))$, where

(8.2) $$\varepsilon(y) = \min\left[\varepsilon(x), f(y, x)\right].$$

(Such y are now labeled and unscanned and x is now labeled and scanned.) Repeat the general step until either the sink t is labeled and unscanned, or until no more labels can be assigned and the sink is unlabeled. In the former case, go to *Routine* B; in the latter case, terminate.

Routine B (flow change). The sink t has been labeled $(y^{\pm}, \varepsilon(t))$. If t is labeled $(y^{+}, \varepsilon(t))$, replace $f(y, t)$ by $f(y, t) + \varepsilon(t)$; if t is labeled $(y^{-}, \varepsilon(t))$, replace $f(t, y)$ by $f(t, y) - \varepsilon(t)$. In either case, next turn attention to node y. In general, if y is labeled $(x^{+}, \varepsilon(y))$, replace $f(x, y)$ by $f(x, y) + \varepsilon(t)$, and if labeled $(x^{-}, \varepsilon(y))$, replace $f(y, x)$ by $f(y, x) - \varepsilon(t)$, and go on to node x. Stop the flow change when the source s is reached, discard the old labels, and go back to *Routine* A.

The labeling process is a systematic search for a flow augmenting path from s to t (Corollary 5.2). Enough information is carried along in the labels so that if the sink is labeled (henceforth we term this case *breakthrough*), the resulting flow change along the path can be made readily. If, on the other hand, *Routine* A ends and the sink has not been labeled (*non-breakthrough*), the flow is maximal and the set of arcs leading from labeled to unlabeled nodes is a minimal cut, since the labeled nodes correspond to the set X defined in the proof of Theorem 5.1.

A main reason underlying the computational efficiency of the labeling process is that once a node is labeled and scanned it can be ignored for the remainder of the process. Labeling a node x corresponds to locating a path from s to x that can be the initial segment of a flow augmenting path. While there may be many such paths from s to x, finding one suffices.

If the flow f is integral and *Routine* A results in breakthrough, then the flow change $\varepsilon(t)$ of *Routine* B, being the minimum of positive integers, is a positive integer. Hence if the computation is initiated with an integral flow, each successive flow is integral. Consequently the algorithm is finite, since the flow value increases by at least one unit with each occurrence of breakthrough; upon termination, a maximal flow has been constructed that is integral. Although this fact is a trivial consequence of the construction, the fact itself is important and will be used time and again in the solution of combinatorial problems. We therefore state it as a theorem.

THEOREM 8.1 (Integrity theorem). *If the capacity function c is integral valued, there exists a maximal flow f that is also integral valued.*

The following numerical example illustrates the use of the labeling method in constructing a maximal flow.

EXAMPLE. Let the given network be that of Fig. 1.1 with arc capacities and initial flow as indicated in Fig. 8.1, the pair $c(x, y), f(x, y)$ being written in that order adjacent to arc (x, y).

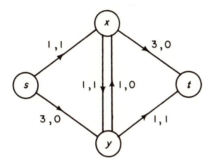

Figure 8.1

Start *Routine* A by assigning s the label $(-, \infty)$, see Fig. 8.2. From s,

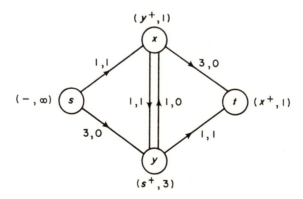

Figure 8.2

label y with $(s^+, \min (3, \infty)) = (s^+, 3)$, thus completing the labeling from s. From y, x can be labeled $(y^+, 1)$ (or $(y^-, 1)$), and is the only unlabeled node

19

that can be labeled from y. Again select a labeled, unscanned node (x is the only such), and continue assigning labels. This time breakthrough occurs: the sink t can be labeled (x^+, 1). This locates a flow augmenting path, found by backtracking from the sink according to the directions given in the labels, along which a flow change of $\varepsilon(t) = 1$ can be made. Here the path is the chain s, y, x, t. The new flow of value 2 is shown in Fig. 8.3.

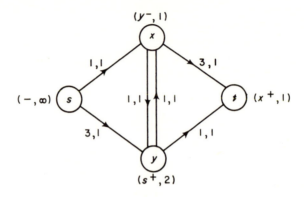

Figure 8.3

Now discard the old labels and repeat the labeling process. This time the labels shown in Fig. 8.3 are obtained. Again breakthrough has resulted and a flow improvement of $\varepsilon(t) = 1$ can be made along the path s, (s, y), y, (x, y), x, (x, t), t, yielding the flow shown in Fig. 8.4.

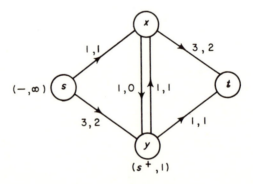

Figure 8.4

Repetition of *Routine* A now results in non-breakthrough, the labeled set of nodes being those shown in Fig. 8.4. Thus the flow of Fig. 8.4 is maximal and a minimal cut consists of the arcs (s, x), (y, x), and (y, t).

Labeling backward from the sink by rules corresponding to (a'), (b') of

§ 5 locates the same cut, and hence by Theorem 5.6 this is the unique minimal cut separating s and t.

We conclude this section with an example indicating that the labeling process might fail to terminate if arc capacities are irrational. Specifically, the example shows that if the process is interpreted broadly enough to permit the selection of any flow augmenting path at each stage of the computation, then finite termination may not occur when arc capacities are irrational.

Before describing this example, we make one definition which will be helpful in the description. If $[N; \mathscr{A}]$ is a network with capacity function c, and if f is a flow from s to t in $[N; \mathscr{A}]$, then $c(x, y) - f(x, y)$ is the *residual capacity* of arc (x, y) with respect to f.

Now consider the recursion

$$a_{n+2} = a_n - a_{n+1}.$$

This recursion has a solution $a_n = r^n$, where $r = (-1 + \sqrt{5})/2 < 1$. Thus the series $\sum_{n=1}^{\infty} a_n$ converges to some sum S. We construct a directed network with four "special arcs"

$$A_1 = (x_1, y_1),$$
$$A_2 = (x_2, y_2),$$
$$A_3 = (x_3, y_3),$$
$$A_4 = (x_4, y_4),$$

and the additional arcs (y_i, y_j), (x_i, y_j), (y_i, x_j), for $i \neq j$, together with source arcs (s, x_i) and sink arcs (y_i, t). The four special arcs have capacities a_0, a_1, a_2, a_2, respectively; all other arcs have capacity S.

Step 1. Find a chain from s to t that includes, from among the special arcs, only A_1, and impose a_0 units of flow in this chain. For example, take the chain s, x_1, y_1, t. (The special arcs now have residual capacities 0, a_1, a_2, a_2, respectively.)

Inductive step. Suppose the special arcs A_1', A_2', A_3', A_4' (some rearrangement of A_1, A_2, A_3, A_4) have residual capacities 0, a_n, a_{n+1}, a_{n+1}. Find a chain from s to t that includes, from among the special arcs, only A_2' and A_3', and impose a_{n+1} additional units of flow along this chain. For example, the chain s, x_2', y_2', x_3', y_3', t will do. (The special arcs now have residual capacities 0, $a_n - a_{n+1} = a_{n+2}$, 0, a_{n+1}.) Next find a path from s to t that contains A_2' as a forward arc, A_1' and A_3' as reverse arcs, the latter being the only reverse arcs of the path, and impose an additional flow of a_{n+2} units along this path. For example, the path s, x_2', y_2', y_1', x_1', y_3', x_3', y_4', t containing the reverse arcs (y_1', x_1'), (y_3', x_3') will do. (The special arcs now have residual capacities a_{n+2}, 0, a_{n+2}, a_{n+1}.)

The inductive step increases the flow value by $a_{n+1} + a_{n+2} = a_n$. Hence no non-special arc is ever required to carry more than $\sum_{n=1}^{\infty} a_n = S$ units

21

of flow in repeating the inductive step. The process converges to a flow having value S, whereas the maximal flow value for this network is $4S$.

9. Lower bounds on arc flows

Although lower bounds of zero have been assumed on all arc flows, there is no real necessity for this assumption in constructing maximal flows. If the conditions

$$(9.1) \qquad\qquad 0 \leqslant f(x, y) \leqslant c(x, y)$$

are replaced by

$$(9.2) \qquad\qquad l(x, y) \leqslant f(x, y) \leqslant c(x, y),$$

where l is a given real valued function defined on arcs of $\mathscr{A}$ that satisfies

$$(9.3) \qquad\qquad 0 \leqslant l(x, y) \leqslant c(x, y),$$

the labeling process can be varied to handle this situation provided one has an initial flow to start the computation. There may be no function f satisfying the equations (2.1) and the inequalities (9.2) (e.g., take $l = c$ in the example of the preceding section), but assuming that these constraints are compatible for a given integral valued l and c, and that an initial f satisfying them has been found, the only change in the labeling rules for constructing a maximal flow is the following. If x has been labeled $(z^\pm, \varepsilon)$, then y may be labeled $[x^-, \min(\varepsilon, f(y, x) - l(y, x))]$ provided $f(y, x) > l(y, x)$.

It is also easy to see that the analogue of Theorem 5.1 becomes

THEOREM 9.1. *If there is a function f satisfying (2.1) and (9.2) for some number v, then the maximal value of v subject to these constraints is equal to the minimum of $c(X, \overline{X}) - l(\overline{X}, X)$ taken over all $X \subseteq N$ with $s \in X$, $t \in \overline{X}$.*

On the other hand, still assuming the existence of a function f satisfying (2.1) and (9.2) for some v, the minimal value of v may be found in a similar way: if x is labeled $(z^\pm, \varepsilon)$ and if $f(x, y) > l(x, y)$, attach the label $[x^-, \min(\varepsilon, f(x, y) - l(x, y))]$ to y; or if $f(y, x) < c(y, x)$, assign y the label $[x^+, \min(\varepsilon, c(y, x) - f(y, x))]$.

Here the analogue of Theorem 5.1 is

THEOREM 9.2. *If there is a function f satisfying (2.1) and (9.2) for some number v, the minimal value of v subject to these constraints is equal to the maximum of $l(X, \overline{X}) - c(\overline{X}, X)$ taken over all $X \subseteq N$ with $s \in X$, $t \in \overline{X}$.*

The questions that still remain are those of determining conditions under which the constraints (2.1) and (9.2) are compatible, and of constructing a function f satisfying them when these conditions hold. We postpone these questions for the moment. They, and similar questions, will be taken up in Chapter II.

22

10. Flows in undirected and mixed networks

Let us suppose that the network is undirected or mixed, and that each arc has a non-negative flow capacity. If the arc (x, y) is undirected with capacity $c(x, y)$, we intepret this to mean that

$$\begin{aligned} f(x, y) &\leqslant c(x, y), \\ f(y, x) &\leqslant c(x, y), \\ f(x, y) \cdot f(y, x) &= 0. \end{aligned}$$

(10.1)

That is, $f(x, y)$ is the flow from x to y in (x, y), and the arc (x, y) has a flow capacity $c(x, y)$ in either direction, but flow is permitted in only one of the two directions.

For example, one might think of a network of city streets, each street having a traffic flow capacity, and ask the question: how should one-way signs be put up on streets not already oriented in order to permit the largest traffic flow from some set of points to another?

At first glance, it might appear that this problem would involve examination of a large number of maximal flow problems obtained by orienting the network in various ways. But a moment's thought shows that the problem can be solved by considering only one directed network: namely, that obtained by replacing each undirected arc with a pair of oppositely directed arcs, each having capacity equal to the old arc. The reason for this is that, given any solution f, v of the flow constraints (2.1) and (2.2), one can produce a solution f', v in which

$$f'(x, y) \cdot f'(y, x) = 0$$

by taking

(10.2) $$f'(x, y) = \max{(0, f(x, y) - f(y, x))}.$$

In words, we can cancel arc flows in opposite directions.

Thus, since it is clear that the maximal flow value for any specific orientation of the given network is no greater than the maximal flow value obtained by replacing each undirected arc by a pair of directed arcs, allowing both orientations for each undirected arc solves the original problem of maximizing v subject to the flow equations (2.1), capacity constraints (2.2) for directed arcs, and constraints (10.1) for undirected arcs.

11. Node capacities and other extensions

Other kinds of inequality constraints in addition to bounds on arc flows can be imposed without altering the character of the maximal flow problem. For instance, suppose that each node x has a flow capacity $k(x) \geqslant 0$, and that it is desired to find a maximal flow from s to t subject to both arc and node capacities.

More explicitly, let us assume that all source arcs are directed from the source and all sink arcs into the sink, and that it is desired to maximize $f(s, N)$ subject to

(11.1) $$f(x, N) - f(N, x) = 0, \qquad x \neq s, t,$$

(11.2) $$0 \leqslant f(x, y) \leqslant c(x, y), \qquad (x, y) \in \mathscr{A},$$

(11.3) $$f(x, N) \leqslant k(x), \qquad x \neq t,$$

(11.4) $$f(N, t) \leqslant k(t).$$

This problem can be reduced to the arc capacity case by a simple device. Define a new network $[N^*; \mathscr{A}^*]$ from $[N; \mathscr{A}]$ as follows. To each $x \in N$ we make correspond two nodes $x', x'' \in N^*$; if $(x, y) \in \mathscr{A}$, then $(x', y'') \in \mathscr{A}^*$; in addition, $(x'', x') \in \mathscr{A}^*$ for each $x \in N$. The (arc) capacity function defined on $\mathscr{A}^*$ is

(11.5) $$c^*(x', y'') = c(x, y), \qquad (x, y) \in \mathscr{A},$$

(11.6) $$c^*(x'', x') = k(x), \qquad x \in N.$$

Thus, for example, if the given network $[N; \mathscr{A}]$ is that of Fig. 11.1, the network $[N^*; \mathscr{A}^*]$ is shown in Fig. 11.2.

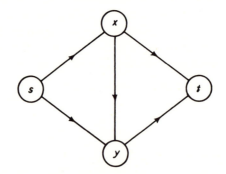

Figure 11.1

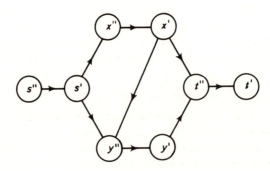

Figure 11.2

In effect, each node x has been split into two parts, a "left" part x'' and a "right" part x', so that all arcs entering x now enter its left part, whereas all arcs leaving x now leave its right part. The capacity $k(x)$ is then imposed as an arc capacity on the new arc leading from the left part of x to its right part.

Thus any function f satisfying (11.1)–(11.4), that is, any flow from s to t in $[N; \mathscr{A}]$ that does not exceed the node capacities, yields an equivalent flow f^* from s'' to t' in $[N^*; \mathscr{A}^*]$ by defining

$$(11.7) \qquad\qquad f^*(x', y'') = f(x, y), \qquad\qquad (x, y) \in \mathscr{A},$$

$$(11.8) \qquad\qquad f^*(x'', x') = f(x, N), \qquad\qquad x \neq t,$$

$$(11.9) \qquad\qquad f^*(t'', t') = f(N, t),$$

and conversely.

If the notion of a disconnecting set is extended to include nodes as well as arcs, the analogue of the max-flow min-cut theorem asserts that the maximal flow value is equal to the capacity of a disconnecting set of nodes and arcs having minimal capacity.

In a similar way, more general kinds of constraints on the flow out of or into node x can be reduced to the case of arc capacities by enlarging the network. For example, suppose that the nodes of the set $A(x)$ are put into subsets

$$(11.10) \qquad\qquad A_1(x), \ldots, A_{m(x)}(x)$$

with the proviso that

$$(11.11) \quad A_i(x) \cap A_j(x) \neq \varnothing \Rightarrow A_i(x) \subseteq A_j(x) \quad \text{or} \quad A_j(x) \subseteq A_i(x),$$

and assume, in addition to the flow equations,

$$(11.12) \qquad\qquad f(x, A_i(x)) \leqslant k_i(x), \qquad\qquad i = 1, \ldots, m(x).$$

Constraints of the form (11.12), under the assumption (11.11), can be handled as indicated schematically in Fig. 11.3 and Fig. 11.4 for a single node x.

Constraints of a similar kind on flow into x can be reduced to arc constraints by enlarging the network in an analogous fashion.

Notice that inequality constraints (11.2), (11.3), (11.4) are a special case of (11.12) and similar constraints on flow into x:

$$(11.13) \qquad\qquad f(B_j(x), x) \leqslant h_j(x), \qquad\qquad j = 1, \ldots, n(x).$$

If we refer to each set $(x, A_i(x))$ and $(B_j(x), x)$ as an elementary set of arcs, and extend the notion of a disconnecting set of arcs to say that a collection $\mathscr{B}$ of elementary sets is a disconnecting collection if each chain from s to t has an arc in common with some elementary set contained in $\mathscr{B}$, it can be shown that the maximal flow value from s to t is equal to the

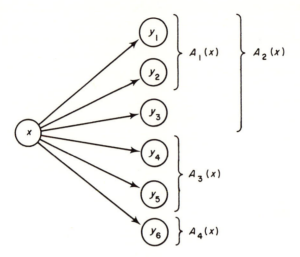

Figure 11.3

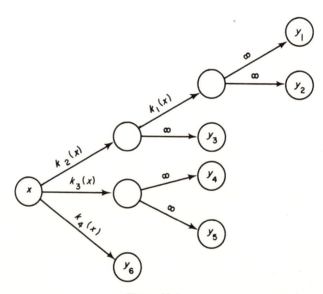

Figure 11.4

minimal blocking capacity (under the assumption (11.11) and a similar assumption on $B_j(x)$).

12. Linear programming and duality principles

The problem of finding a maximal flow through a network, whether stated in node-arc or in arc-chain form, is one of extremizing a linear

26

function subject to linear equations and linear inequalities. Such a problem is called a linear programming problem. There are various known methods of computing answers to linear programs. The method that is in general use is G. B. Dantzig's simplex algorithm, around which a sizeable literature has already grown up. It is not our purpose here to discuss the theory of linear inequalities or algorithms for solving general linear programs, since this book is devoted, for the most part, to special kinds of linear programs that arise in transportation, communication, or certain kinds of combinatorial problems, and to a presentation of special algorithms for solving these linear programs. We would be negligent, however, if some mention were not made of linear programming duality principles in connection with these problems.

Associated with every linear programming problem in variables $w_1, \ldots, w_n$:

$$a_{11}w_1 + \ldots + a_{1l}w_l + a_{1,l+1}w_{l+1} + \ldots + a_{1n}w_n = b_1$$
$$\ldots$$
$$a_{k1}w_1 + \ldots + a_{kl}w_l + a_{k,l+1}w_{l+1} + \ldots + a_{kn}w_n = b_k$$

(12.1)

$$a_{k+1,1}w_1 + \ldots + a_{k+1,l}w_l + a_{k+1,l+1}w_{l+1} + \ldots + a_{k+1,n}w_n \leqslant b_{k+1}$$
$$\ldots$$
$$a_{m1}w_1 + \ldots + a_{ml}w_l + a_{m,l+1}w_{l+1} + \ldots + a_{mn}w_n \leqslant b_m$$

(12.2) $\qquad w_1, \ldots, w_l$ unrestricted in sign; $w_{l+1}, \ldots, w_n \geqslant 0$

(12.3) $\qquad \qquad$ maximize $c_1w_1 + \ldots + c_nw_n$

is a *dual program* obtained by assigning multipliers $\lambda_1, \ldots, \lambda_m$ to the individual constraints of (12.1) and forming the program

$$a_{11}\lambda_1 + \ldots + a_{k1}\lambda_k + a_{k+1,1}\lambda_{k+1} + \ldots + a_{m1}\lambda_m = c_1$$
$$\ldots$$
$$a_{1l}\lambda_1 + \ldots + a_{kl}\lambda_k + a_{k+1,l}\lambda_{k+1} + \ldots + a_{ml}\lambda_m = c_l$$

(12.4)

$$a_{1,l+1}\lambda_1 + \ldots + a_{k,l+1}\lambda_k + a_{k+1,l+1}\lambda_{k+1} + \ldots + a_{m,l+1}\lambda_m \geqslant c_{l+1}$$
$$\ldots$$
$$a_{1n}\lambda_1 + \ldots + a_{kn}\lambda_k + a_{k+1,n}\lambda_{k+1} + \ldots + a_{mn}\lambda_m \geqslant c_m$$

(12.5) $\qquad \lambda_1, \ldots, \lambda_k$ unrestricted in sign; $\lambda_{k+1}, \ldots, \lambda_m \geqslant 0$

(12.6) $\qquad \qquad$ minimize $b_1\lambda_1 + \ldots + b_m\lambda_m$.

Here the a_{ij}, b_i, and c_j are given real numbers.

The matrix of the constraints (12.4) is the transpose of that of (12.1). Equalities of (12.4) correspond to unrestricted variables $w_1, \ldots, w_l$, and inequalities to non-negative variables $w_{l+1}, \ldots, w_n$. The multipliers or dual variables $\lambda_1, \ldots, \lambda_k$ that correspond to equations of (12.1) are unrestricted in sign, whereas $\lambda_{k+1}, \ldots, \lambda_m$, corresponding to inequalities of (12.1), are non-negative.

Observe that if the dual problem (12.4), (12.5), and (12.6) is written in the form of the *primal problem* (12.1), (12.2), and (12.3), by multiplying each of the constraints of (12.4) by -1 and maximizing $-\sum b_i \lambda_i$, then the dual of (12.4), (12.5), (12.6) is (12.1), (12.2), (12.3). In other words, the dual of the dual is the primal.

The constraints of the primal problem are said to be *feasible* if there is a vector $w = (w_1, \ldots, w_n)$ satisfying them; w is then called a *feasible vector*, and the primal problem is termed *feasible*. A feasible vector w that maximizes the linear form $\sum c_j w_j$ is called *optimal*. Analogous language is used for the dual problem.

Thus a linear programming problem either has

(a) optimal (and hence feasible) vectors;
(b) feasible vectors, but no optimal vector;
(c) no feasible vectors.

The fundamental duality theorem of linear programming [9] relates the way these situations can occur in a pair of dual programs, and asserts equality between the maximum in the primal and the minimum in the dual: if case (a) holds for the primal, then (a) holds for its dual and the maximum value of $\sum c_j w_j$ is equal to the minimum value of $\sum b_i \lambda_i$; if (b) holds for the primal, then (c) holds for the dual; if (c) holds for the primal, either (b) or (c) is valid for the dual.

That the maximum value of $\sum c_j w_j$ is no greater than the minimum of $\sum b_i \lambda_i$ if both primal and dual have feasible vectors is easily seen. Letting w and λ be feasible in their respective programs, it follows that

$$(12.7) \qquad \sum_j c_j w_j \leqslant \sum_j \sum_i \lambda_i a_{ij} w_j,$$

since unrestricted variables w_j correspond to equations $\sum_i \lambda_i a_{ij} = c_j$ and non-negative variables w_j to inequalities $\sum_i \lambda_i a_{ij} \geqslant c_j$.

Thus equality holds in (12.7) if and only if

$$(12.8) \qquad \sum_i \lambda_i a_{ij} > c_j \Rightarrow w_j = 0.$$

Similarly,

$$(12.9) \qquad \sum_j \sum_i \lambda_i a_{ij} w_j \leqslant \sum_i b_i \lambda_i,$$

since the λ_i that are unrestricted in sign correspond to equations $\sum_j a_{ij} w_j = b_i$, whereas non-negative λ_i correspond to inequalities $\sum_j a_{ij} w_j \leqslant b_i$.

28

Thus, equality holds in (12.9) if and only if

$$(12.10) \qquad \lambda_i > 0 \Rightarrow \sum_j a_{ij} w_j = b_i.$$

Consequently

$$(12.11) \qquad \sum c_j w_j \leqslant \sum b_i \lambda_i,$$

equality holding if and only if (12.8) and (12.10) are valid. The major content of the duality theorem is the assertion that if case (a) holds for the primal, it also holds for the dual, and that there are then feasible solutions to primal and dual problems that satisfy the optimality criteria (12.8) and (12.10).

Our purpose in giving this sketchy résumé of linear programming duality theory is twofold. First, we shall note that the max-flow min-cut theorem provides a proof of the duality theorem for the special case of maximal flow problems. Second, although the algorithms to be presented subsequently do not require appeal to the duality theorem, they were motivated by duality considerations, and we want to feel free to invoke such considerations where convenient.

If we take the constraints of the maximal flow problem in the node-arc form and assign multipliers $\pi(x)$ to the equations (2.1), multipliers $\gamma(x, y)$ to the capacity inequalities (2.2), then, recalling that the coefficient matrix of the equations is (apart from the column corresponding to the variable v) the node-arc incidence matrix of the network, it follows that the dual has constraints

$$(12.12) \qquad \begin{aligned} -\pi(s) + \pi(t) &\geqslant 1, \\ \pi(x) - \pi(y) + \gamma(x, y) &\geqslant 0, &\text{all } (x, y), \\ \gamma(x, y) &\geqslant 0, &\text{all } (x, y), \end{aligned}$$

subject to which the form

$$(12.13) \qquad \sum_{\mathscr{A}} c(x, y)\gamma(x, y)$$

is to be minimized. In (12.12), the first constraint comes from the v-column of the primal problem, the second from the (x, y)-column. The dual variables $\pi(x)$ are unrestricted in sign since they correspond to equations, whereas the variables $\gamma(x, y)$ correspond to inequalities and are consequently non-negative.

If $(X, \overline{X})$ is a minimal cut separating s and t, it can be checked that an optimal solution to the dual problem is provided by taking

$$(12.14) \qquad \pi(x) = \begin{cases} 0 & \text{for } x \in X, \\ 1 & \text{for } x \in \overline{X}, \end{cases}$$

$$(12.15) \qquad \gamma(x, y) = \begin{cases} 1 & \text{for } (x, y) \in (X, \overline{X}), \\ 0 & \text{otherwise.} \end{cases}$$

29

This follows since (12.14) and (12.15) define a feasible solution to the dual that produces equality between the primal form v and dual form (12.13). Or one can check the optimality properties (12.8) and (12.10).

In particular, the dual of the maximal flow problem always has an integral solution. It can be shown, in fact, that all extreme points of the convex polyhedral set defined by setting $\pi(s) = 0$ in (12.12), which corresponds to dropping the (redundant) source equation in the primal problem, are of the form given in (12.14) and (12.15) for some X with $s \in X$. Using this fact, the max-flow min-cut theorem can be deduced from the duality theorem [2].

13. Maximal flow value as a function of two arc capacities

For a given network $[N; \mathscr{A}]$ with specified sources S and sinks T, the value $\bar{v}$ of a maximal flow from S to T is solely a function of the individual arc capacities. Indeed, if $\mathscr{A} = \{A_1, A_2, \ldots, A_m\}$ and A_i has capacity $c(A_i)$, we know that

$$(13.1) \qquad \bar{v} = \min_{\mathscr{C} \subseteq \mathscr{A}} \sum_{A_i \in \mathscr{C}} c(A_i),$$

the minimum being taken over all cuts $\mathscr{C}$ separating S and T. The theorems and proofs of this section provide insight into the behavior of $\bar{v}$ considered as a function of two arc capacities, everything else being held fixed. Both theorems and proofs are due to Shapley [16].

It will be convenient to allow infinite capacities for the two arcs in question, and hence infinite $\bar{v}$. However, the capacities of other arcs are assumed finite.

Let $\bar{v}_i(\xi)$ denote the maximal flow value when the capacity $c(A_i)$ has been replaced by the non-negative variable ξ. Similarly, $\bar{v}_{ij}(\xi, \eta)$ denotes the maximal flow value when $c(A_i)$ and $c(A_j)$ have been replaced by non-negative variables ξ and η. It is a consequence of (13.1) that

$$(13.2) \qquad \bar{v}_i(\xi) = \min[\bar{v}_i(0) + \xi, \bar{v}_i(\infty)].$$

In more detail, if ξ is less than the *critical capacity*

$$(13.3) \qquad \xi^* = \bar{v}_i(\infty) - \bar{v}_i(0),$$

the arc A_i is a member of every minimal cut, whereas for $\xi > \xi^*$, the arc A_i is in no minimal cut. Here ξ^* may be either zero or infinite. If the critical capacity ξ^* is strictly positive, and if $c(A_i) = \xi^*$, there is a minimal cut containing A_i and a minimal cut not containing A_i.

Two applications of (13.2) yield

$$(13.4) \quad \bar{v}_{ij}(\xi, \eta) = \min[\bar{v}_{ij}(0, 0) + \xi + \eta, \ \bar{v}_{ij}(0, \infty) + \xi,$$
$$\bar{v}_{ij}(\infty, 0) + \eta, \bar{v}_{ij}(\infty, \infty)].$$

Thus the piecewise linear function $\bar{v}_{ij}(\xi, \eta)$ divides the non-negative quadrant of the ξ, η plane into at most four open convex regions in each of which it is linear, together with certain boundary lines and vertices. We label these regions R_{11}, R_{10}, R_{01}, R_{00}, respectively: R_{11} is the region in which the minimum in (13.4) is assumed uniquely by $\bar{v}_{ij}(0, 0) + \xi + \eta$, R_{10} the region in which the minimum is assumed uniquely by $\bar{v}_{ij}(0, \infty) + \xi$, and so on. Thus the subscripts identifying the region are the values of the partial derivatives $\dfrac{\partial \bar{v}_{ij}}{\partial \xi}, \dfrac{\partial \bar{v}_{ij}}{\partial \eta}$ in that region. Notice that for any point of R_{11}, both arcs A_i and A_j are in every minimal cut; in R_{10}, A_i is in every minimal cut while A_j is in no minimal cut; in R_{01}, A_i is in no minimal cut and A_j is in every minimal cut; in R_{00}, neither A_i nor A_j is in any minimal cut.

The common boundaries of each pair of regions appear as in Fig. 13.1:

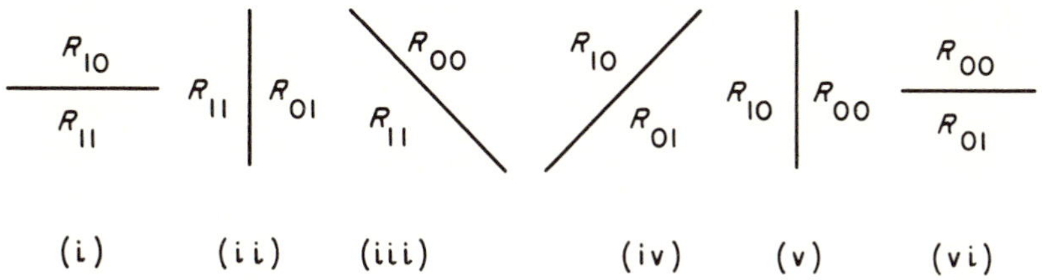

Figure 13.1

The equations of these boundary lines are respectively

$$(13.5) \qquad \eta = \bar{v}_{ij}(0, \infty) - \bar{v}_{ij}(0, 0) \quad (R_{10} - R_{11})$$

$$(13.6) \qquad \xi = \bar{v}_{ij}(\infty, 0) - \bar{v}_{ij}(0, 0) \quad (R_{11} - R_{01})$$

$$(13.7) \qquad \xi + \eta = \bar{v}_{ij}(\infty, \infty) - \bar{v}_{ij}(0, 0) \quad (R_{11} - R_{00})$$

$$(13.8) \qquad \xi - \eta = \bar{v}_{ij}(\infty, 0) - \bar{v}_{ij}(0, \infty) \quad (R_{10} - R_{01})$$

$$(13.9) \qquad \xi = \bar{v}_{ij}(\infty, \infty) - \bar{v}_{ij}(0, \infty) \quad (R_{10} - R_{00})$$

$$(13.10) \qquad \eta = \bar{v}_{ij}(\infty, \infty) - \bar{v}_{ij}(\infty, 0) \quad (R_{00} - R_{01}).$$

Here $\bar{v}_{ij}(0, \infty) = \infty$ means that region R_{10} is empty, $\bar{v}_{ij}(\infty, 0) = \infty$ means that R_{01} is empty, and $\bar{v}_{ij}(\infty, \infty) = \infty$ means that R_{00} is empty.

In order to determine the different ways in which the non-negative quadrant of the ξ, η plane can be partitioned by the four regions, a case classification can be made using

$$(13.11) \qquad p_{ij} = \bar{v}_{ij}(\infty, \infty) - \bar{v}_{ij}(0, \infty) - \bar{v}_{ij}(\infty, 0) + \bar{v}_{ij}(0, 0)$$

as follows:

    (a) $p_{ij} > 0$ (including $p_{ij} = \infty$),
    (b) $p_{ij} = 0$ or p_{ij} indeterminate $(\infty - \infty)$,
    (c) $p_{ij} < 0$.

Using (13.5)–(13.10), it follows that if all four regions are present in each case, the resulting configurations for the ξ, η non-negative quadrant then appear as in Fig. 13.2:

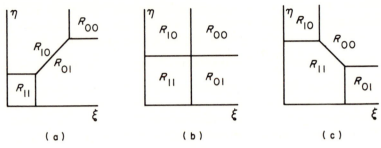

Figure 13.2

Moreover, if $p_{ij} = \infty$, the configuration is a degenerate form of Fig. 13.2(a) in which R_{00} does not appear, while if p_{ij} is indeterminate, various degenerate forms of Fig. 13.2(b) occur. Of course, other kinds of degeneracy may be present, e.g., R_{10} may be empty in Fig. 13.2(c) by virtue of $\bar{v}_{ij}(\infty, \infty) - \bar{v}_{ij}(0, \infty) = 0$, and so on. But the configurations of Fig. 13.2 are exclusive and comprehend all possibilities. Notice that there is never more than one diagonal boundary segment, that is, an $R_{11} - R_{00}$ contact precludes an $R_{10} - R_{01}$ contact, and that in cases (a) and (c), a diagonal segment is always present. For future reference, we also note that a point (ξ^*, η^*) on a diagonal segment is critical in the following sense: if $c(A_i)$ is fixed at ξ^*, then η^* is the critical capacity of A_j, whereas if $c(A_j)$ is fixed at η^*, then ξ^* is the critical capacity of A_i. Thus at such a point (ξ^*, η^*) with $\xi^* > 0, \eta^* > 0$, there is a minimal cut containing A_i, a minimal cut not containing A_i, and similarly for A_j.

The foregoing case classification provides the background for a general statement about the difference quotient

$$(13.12) \quad q_{ij} = \frac{\bar{v}_{ij}(\xi + h, \eta + k) - \bar{v}_{ij}(\xi + h, \eta) - \bar{v}_{ij}(\xi, \eta + h) + \bar{v}_{ij}(\xi, \eta)}{hk}$$

for the function $\bar{v}_{ij}(\xi, \eta)$. Here q_{ij} is of course a function of h and k as well as ξ and η, and is well defined only if $\xi + h \geqslant 0, \eta + k \geqslant 0$, and $hk \neq 0$.

THEOREM 13.1. *For all rectangles (ξ, η), $(\xi + h, \eta)$, $(\xi, \eta + k)$, $(\xi + h, \eta + k)$ in the ξ, η non-negative quadrant, the difference quotient q_{ij} is of one sign.*

PROOF. Assume without loss of generality that $h > 0, k > 0$, and consider the described rectangle. It cannot enclose more than one diagonal piece from the boundary configuration. If it encloses none, then $q_{ij} = 0$. If the piece enclosed has positive slope, then $q_{ij} > 0$ (in fact, q_{ij} is equal to the length of the intercepted diagonal divided by $hk\sqrt{2}$). On the other hand, if the piece enclosed has negative slope, then $q_{ij} < 0$.

The following corollary, which relates the sign of q_{ij} to that of the constant p_{ij} defined by (13.11), is immediate.

COROLLARY 13.2.

(a) *If $p_{ij} > 0$ (including $p_{ij} = \infty$), then q_{ij} is sometimes > 0, and never < 0;*

(b) *if $p_{ij} = 0$ or if p_{ij} is indeterminate $(\infty - \infty)$, then q_{ij} is identically zero;*

(c) *if $p_{ij} < 0$, then q_{ij} is sometimes < 0, and never > 0.*

Theorem 13.1 can be verbalized in a somewhat more intuitive way. Roughly speaking, a positive q_{ij} means that the arcs A_i and A_j complement or reinforce each other, whereas a negative q_{ij} means that they compete or interfere with each other. Thus the theorem asserts that any pair of arcs in a network (having fixed capacities for all other arcs) consistently reinforce or interfere with each other. In general, the manner in which two arcs interact depends on the capacities of the other arcs, as well as the relative positions of the two arcs in the network. For example, consider the network of Fig. 13.3:

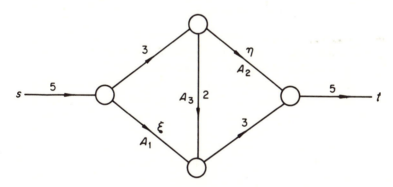

Figure 13.3

Here $p_{12} = 1$, but removal of the arc A_3 (or reducing its capacity to zero) yields $p_{12} = -1$. However, in certain cases, the interaction-type of a pair of arcs is determined solely by their relative positions, independently of the capacity values, as the following theorem and corollary show.

THEOREM 13.3.

(i) *If the terminal node of A_i is the initial node of A_j, then $q_{ij} \geqslant 0$.*

(ii) *If A_i and A_j have the same initial node, then $q_{ij} \leqslant 0$.*

(iii) *If the initial node of A_i is a source, and the terminal node of A_j is a sink, then $q_{ij} \geqslant 0$.*

Proof.

(i) Consider a minimal cut $\mathscr{C} = (X, \overline{X})$ in the network. If the common node x of A_i and A_j is in X, then A_i is not in $\mathscr{C}$. If x is in $\overline{X}$, then A_j is not in $\mathscr{C}$. Thus R_{11}, in which both A_i and A_j belong to every minimal cut, is empty. Hence $q_{ij} \geqslant 0$.

(ii) Ignoring the trivial case in which no diagonal segment appears in the configuration, let (ξ^*, η^*) be an arbitrary point on the diagonal having positive coordinates. Then (ξ^*, η^*) is critical, and hence there is a minimal cut $\mathscr{C}_1 = (X_1, \overline{X}_1)$ containing A_i and a minimal cut $\mathscr{C}_2 = (X_2, \overline{X}_2)$ containing A_j. Thus, since A_i and A_j have the same initial node, the minimal cut $\mathscr{C} = (X_1 \cap X_2, \overline{X}_1 \cup \overline{X}_2)$ contains both A_i and A_j. The capacity of $\mathscr{C}$ corresponding to the point (ξ^*, η^*) is of course $\bar{v}_{ij}(\xi^*, \eta^*)$, and consequently the capacity of $\mathscr{C}$ corresponding to the variable point (ξ, η) is $\bar{v}_{ij}(\xi^*, \eta^*) + \xi - \xi^* + \eta - \eta^*$. Thus

$$\bar{v}_{ij}(\xi, \eta) \leqslant \bar{v}_{ij}(\xi^*, \eta^*) + \xi - \xi^* + \eta - \eta^*,$$

and in particular,

$$\bar{v}_{ij}(0, 0) \leqslant \bar{v}_{ij}(\xi^*, \eta^*) - \xi^* - \eta^*.$$

By (13.4) equality holds here, and so

$$\bar{v}_{ij}(\xi^*, \eta^*) = \bar{v}_{ij}(0, 0) + \xi^* + \eta^*.$$

It follows that (ξ^*, η^*) is on the boundary of R_{11}. Since (ξ^*, η^*) was an arbitrary point on the diagonal having positive coordinates, the boundary configuration is that of Fig. 13.2(c), and hence $q_{ij} \leqslant 0$.

(iii) The proof here is similar to that of (ii). Again we may ignore the trivial case corresponding to Fig. 13.2(b), and select a critical point (ξ^*, η^*) on the boundary segment having positive coordinates. Hence there is a minimal cut $\mathscr{C}_1 = (X_1, \overline{X}_1)$ containing A_i and a minimal cut $\mathscr{C}_2 = (X_2, \overline{X}_2)$ not containing A_j. It follows that the minimal cut $\mathscr{C} = (X_1 \cap X_2, \overline{X}_1 \cup \overline{X}_2)$ contains A_i but not A_j. The capacity of $\mathscr{C}$ corresponding to (ξ^*, η^*) is $\bar{v}_{ij}(\xi^*, \eta^*)$, and hence the capacity of $\mathscr{C}$ corresponding to the variable point (ξ, η) is $\bar{v}_{ij}(\xi^*, \eta^*) + \xi - \xi^*$. Thus

$$\bar{v}_{ij}(\xi, \eta) \leqslant \bar{v}_{ij}(\xi^*, \eta^*) + \xi - \xi^*,$$

and in particular

$$\bar{v}_{ij}(0, \infty) \leqslant \bar{v}_{ij}(\xi^*, \eta^*) - \xi^*.$$

Again equality must hold here, and so (ξ^*, η^*) is on the boundary of R_{10}. Hence $q_{ij} \geqslant 0$.

COROLLARY 13.4. *If A_i and A_j have the same terminal node, then $q_{ij} \leqslant 0$. If the initial nodes of A_i and A_j are both sources, then $q_{ij} \leqslant 0$. If the terminal nodes of A_i and A_j are both sinks, then $q_{ij} \leqslant 0$.*

PROOF. The first statement follows from the theorem by reversing all arc orientations and interchanging the roles of sources and sinks. The second statement can be proved in a way exactly analogous to the proof of part (ii) of the theorem. The third statement follows from the second by reversing the network.

References

1. G. B. Dantzig, "Application of the Simplex Method to a Transportation Problem," *Activity Analysis of Production and Allocation*, Cowles Commission Monograph 13, Wiley, 1951, 359–373.

2. ———— and D. R. Fulkerson, "On the Max-Flow Min-Cut Theorem of Networks," *Linear Inequalities and Related Systems*, Annals of Mathematics Study 38, Princeton University Press, 1956, 215–221.

3. P. Elias, A. Feinstein, and C. E. Shannon, "Note on Maximum Flow Through a Network," *I.R.E. Trans. on Inform. Theory*, IT-2 (1956), 117–119.

4. L. R. Ford, Jr., and D. R. Fulkerson, "Maximal Flow Through A Network," *Canad. J. Math.* 8 (1956), 399–404.

5. ————, "A Simple Algorithm for Finding Maximal Network Flows and an Application to the Hitchcock Problem," *Canad. J. Math.* 9 (1957), 210–218.

6. ————, "A Suggested Computation for Maximal Multi-Commodity Network Flows," *Management Sci.* 5 (1958), 97–101.

7. D. R. Fulkerson and G. B. Dantzig, "Computation of Maximal Flows in Networks," *Naval Res. Logist. Quart.* 2 (1955), 277–283.

8. D. Gale, "The Basic Theorems of Real Linear Equations, Inequalities, Linear Programming, and Game Theory," *Naval Res. Logist. Quart.* 3 (1956), 193–200.

9. D. Gale, H. W. Kuhn, and A. W. Tucker, "Linear Programming and the Theory of Games," *Activity Analysis of Production and Allocation*, Cowles Commission Monograph 13, Wiley, 1951, 317–328.

10. A. J. Goldman and A. W. Tucker, "Theory of Linear Programming," *Linear Inequalities and Related Systems*, Annals of Mathematics Study 38, Princeton University Press, 1956, 53–97.

11. T. E. Harris and F. S. Ross, "Fundamentals of a Method for Evaluating Rail Net Capacities," (U) The RAND Corporation, Research Memorandum RM-1573, October 24, 1956 (Secret).

12. A. J. Hoffman and J. B. Kruskal, Jr., "Integral Boundary Points of Convex Polyhedra," *Linear Inequalities and Related Systems*, Annals of Mathematics Study 38, Princeton University Press, 1956, 223–246.

13. D. König, *Theorie der Endlichen und Unendlichen Graphen*, Chelsea Publishing Co., New York, 1936.

14. J. T. Robacker, "On Network Theory," The RAND Corporation, Research Memorandum RM-1498, May 26, 1955.

15. ————, "Min-Max Theorems on Shortest Chains and Disjunct Cuts of a Network," The RAND Corporation, Research Memorandum RM-1660, January 12, 1956.

16. L. S. Shapley, "On Network Flow Functions," The RAND Corporation, Research Memorandum RM-2338, March 16, 1959.

CHAPTER II

FEASIBILITY THEOREMS AND COMBINATORIAL APPLICATIONS

Introduction

The first part of this chapter develops several theorems which give necessary and sufficient conditions for the existence of network flows that satisfy additional linear inequalities of various kinds. Adopting the linear programming terminology introduced in the first chapter, we call these feasibility theorems. Typical of such are a supply-demand theorem (§ 1) due to Gale [11], that states conditions for the existence of a flow satisfying "demands" at certain nodes from "supplies" at others, and a circulation theorem (§ 3) due to Hoffman [17] that gives conditions for the existence of a circulatory flow in a network in which arc flows are subject to both lower and upper bounds. In addition to these and variants of them, one other useful feasibility theorem is presented in § 2.

Proofs of each of these theorems can be made to rely on the max-flow min-cut theorem. (It is true, conversely, that each implies the max-flow min-cut theorem.) As a consequence, it will follow from the integrity theorem that if the additional constraints are integral, e.g., if the supply and demand functions in the supply-demand theorem are integral valued, or if the lower and upper bound functions for the circulation theorem are integral valued, then integral feasible flows exist provided there are any feasible flows. Using this fact, various combinatorial problems that have received attention in the mathematical literature can be posed and solved in terms of network flow. The remainder of the chapter illustrates this method of attack on a number of such problems.

1. A supply-demand theorem

Let $[N; \mathscr{A}]$ be an arbitrary network with capacity function c, and suppose that N is partitioned into sources S, intermediate nodes R, and sinks T (as in I.7). Associate with each $x \in S$ a non-negative number $a(x)$, to be thought of as the *supply* of some commodity at x, and with each $x \in T$ a non-negative number $b(x)$, the *demand* for the commodity at x. We are interested in the question: under what conditions can the demands at the sinks be fulfilled from the supplies at the sources, that is, when are the constraints

$$f(x, N) - f(N, x) \leqslant a(x), \qquad\qquad x \in S,$$

$$f(x, N) - f(N, x) = 0, \qquad\qquad x \in R,$$

$$f(N, x) - f(x, N) \geqslant b(x), \qquad\qquad x \in T,$$

$$0 \leqslant f(x, y) \leqslant c(x, y), \qquad\qquad (x, y) \in \mathscr{A},$$

feasible?

For example, consider the undirected network of Fig. 1.1 having arc capacities as indicated, and suppose $a(1) = 7, a(2) = 2, b(7) = 1, b(8) = 8$. (Formally, we take $\mathscr{A}$ to consist of ordered pairs, so that the undirected arc (x, y) is replaced by the pair of directed arcs (x, y) and (y, x), each having the given capacity.)

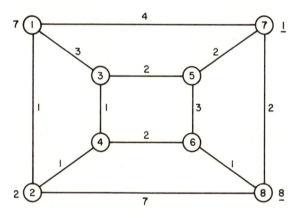

Figure 1.1

A flow that almost succeeds in meeting the demands from the supplies is shown in Fig. 1.2, in which the second numbers on arcs represent the amounts of flow, and arrows denote flow directions.

Is there a feasible flow for this problem, or does Fig. 1.2 represent the best one can do? To answer this question, look at the subset of nodes $\bar{X} = \{2, 8\}$. If the problem is feasible, it must be possible to send into $\bar{X}$ a total amount that is at least equal to the excess of demand over supply for $\bar{X}$, here $b(8) - a(2) = 6$. But the arcs leading into $\bar{X}$ have capacity sum 5. Thus it is not possible to fulfill the demand at node 8, and the problem is infeasible.

The following theorem, due to Gale [11], gives necessary and sufficient conditions for the supply-demand constraints to be feasible.

37

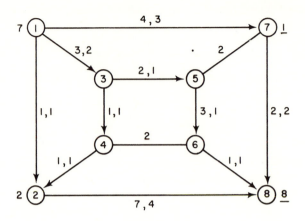

Figure 1.2

THEOREM 1.1. *The constraints*

$$(1.1) \qquad\qquad f(x, N) - f(N, x) \leqslant a(x), \qquad\qquad x \in S,$$

$$(1.2) \qquad\qquad f(x, N) - f(N, x) = 0, \qquad\qquad x \in R,$$

$$(1.3) \qquad\qquad f(N, x) - f(x, N) \geqslant b(x), \qquad\qquad x \in T,$$

$$(1.4) \qquad\qquad 0 \leqslant f(x, y) \leqslant c(x, y), \qquad\qquad (x, y) \in \mathscr{A},$$

where $a(x) \geqslant 0$, $b(x) \geqslant 0$, *are feasible if and only if*

$$(1.5) \qquad\qquad b(T \cap \overline{X}) - a(S \cap \overline{X}) \leqslant c(X, \overline{X})$$

holds for every subset $X \subseteq N$.

Interpreting a supply as a negative demand, condition (1.5) is the statement that the net demand over any subset $\overline{X}$ of N cannot exceed the capacity of the arcs leading into $\overline{X}$. The main content of the theorem is the assertion that if this condition is satisfied for all subsets of N, then the problem is feasible.

PROOF. If there is an f satisfying (1.1)–(1.3), it follows immediately by summing these equations and inequalities over $x \in \overline{X}$ that

$$b(T \cap \overline{X}) - a(S \cap \overline{X}) \leqslant f(N, \overline{X}) - f(\overline{X}, N).$$

Writing $N = X \cup \overline{X}$ gives

$$b(T \cap \overline{X}) - a(S \cap \overline{X}) \leqslant f(X, \overline{X}) - f(\overline{X}, X).$$

If f satisfies (1.4), this last inequality implies

$$b(T \cap \overline{X}) - a(S \cap \overline{X}) \leqslant c(X, \overline{X}).$$

38

§1. A SUPPLY-DEMAND THEOREM

To prove the sufficiency of (1.5), extend the network $[N; \mathscr{A}]$ to a new network $[N^*; \mathscr{A}^*]$ by adjoining a fictitious source s, sink t, and the arcs (s, S), (T, t). The capacity function on $\mathscr{A}^*$ is defined by

$$
\begin{aligned}
c^*(s, x) &= a(x), & x \in S, \\
c^*(x, t) &= b(x), & x \in T, \\
c^*(x, y) &= c(x, y), & (x, y) \in \mathscr{A}.
\end{aligned}
$$

The assumption (1.5) for all $X \subseteq N$ is tantamount to the statement that the cut (T, t) separating s and t is minimal in $[N^*; \mathscr{A}^*]$. To see this, let $(X^*, \overline{X}^*)$ be any cut separating s and t. Defining $X = X^* - s$, $\overline{X} = \overline{X}^* - t$, we have

$$
\begin{aligned}
c^*(X^*, \overline{X}^*) - c^*(T, t) &= c^*(X, t) + c^*(s, \overline{X}) + c^*(X, \overline{X}) - c^*(T, t) \\
&= b(T \cap X) + a(S \cap \overline{X}) + c(X, \overline{X}) - b(T) \\
&= -b(T \cap \overline{X}) + a(S \cap \overline{X}) + c(X, \overline{X}).
\end{aligned}
$$

Thus $c^*(X^*, \overline{X}^*) \geqslant c^*(T, t)$ for all cuts $(X^*, \overline{X}^*)$ separating s and t if and only if (1.5) holds for all $X \subseteq N$.

It follows from the max-flow min-cut theorem that (1.5) implies the existence of a flow f^* from s to t in $[N^*; \mathscr{A}^*]$ that saturates all arcs of (T, t). The restriction f of f^* to $\mathscr{A}$ clearly satisfies (1.2) and (1.4); f also satisfies (1.1) and (1.3), since

$$
a(x) \geqslant f^*(s, x) = f^*(x, N) - f^*(N, x) = f(x, N) - f(N, x),
$$

$$
b(x) = f^*(x, t) = f^*(N, x) - f^*(x, N) = f(N, x) - f(x, N),
$$

for x in S and T, respectively.

This completes the proof of Theorem 1.1.

Going back to the example of Fig. 1.1, we see from the labeling process that a minimal cut in the enlarged network consists of the arcs $(s, 2)$, $(1, 2)$, $(4, 2)$, $(6, 8)$, $(7, 8)$, $(7, t)$ having capacity sum 8, which is less than the total demand 9. Hence the problem is infeasible and a partition $(X, \overline{X})$ of N has been found for which (1.5) fails, namely $\overline{X} = \{2, 8\}$, as was noted earlier. In general, if one is interested in checking the feasibility of a given supply-demand network, the most efficient method is to use the labeling process to solve the equivalent maximal flow problem in the enlarged network, rather than to check the condition (1.5) for all subsets of N. If the problem is infeasible, a violation of (1.5) will be located at the conclusion of the computation by taking X and $\overline{X}$ to be the labeled and unlabeled nodes of N, respectively.

The principal tool used in the proof of Theorem 1.1 was the max-flow min-cut theorem. It can be seen, on the other hand, that Theorem 1.1 implies the max-flow min-cut theorem: one places a demand at the sink t

equal to the minimal cut capacity, an infinite supply at the source s, and lets the other nodes be members of R.

There is another formulation of Theorem 1.1 which is useful [11].

COROLLARY 1.2. *The constraints* (1.1)–(1.4) *are feasible if and only if, for every set* $T' \subseteq T$, *there is a flow* $f_{T'}$ *satisfying* (1.1), (1.2), (1.4), *and*

$$(1.6) \qquad f_{T'}(N, T') - f_{T'}(T', N) \geqslant b(T').$$

PROOF. The necessity is obvious. Sufficiency asserts that if, corresponding to every subset of sinks, there is a flow that satisfies the aggregate demand of the subset without exceeding the supply limitations at each source, then there is a flow that meets all the individual demands.

To prove sufficiency, let $X, \bar{X}$ be any partition of N and define sets

$$S' = S \cap \bar{X}, R' = R \cap \bar{X}, T' = T \cap \bar{X}.$$

Since $f_{T'}$ satisfies (1.1), (1.2), and (1.6), it follows that

$$-a(S') \leqslant f_{T'}(N, S') - f_{T'}(S', N),$$
$$0 = f_{T'}(N, R') - f_{T'}(R', N),$$
$$b(T') \leqslant f_{T'}(N, T') - f_{T'}(T', N).$$

Hence adding and using (1.4),

$$b(T') - a(S') \leqslant f_{T'}(N, \bar{X}) - f_{T'}(\bar{X}, N) = f_{T'}(X, \bar{X}) - f_{T'}(\bar{X}, X)$$
$$\leqslant c(X, \bar{X}).$$

Thus condition (1.5) is fulfilled for all $X \subseteq N$, and the constraints are feasible by Theorem 1.1.

A similar proof shows that the supply-demand constraints are feasible if and only if, for every subset S' of sources, there is a flow $f_{S'}$ satisfying (1.2), (1.3), (1.4) and

$$f_{S'}(S', N) - f_{S'}(N, S') \leqslant a(S').$$

In other words, if corresponding to each subset of sources, there is a flow that satisfies all individual demands without exceeding the aggregate supply of the subset (the supply at sources outside the subset being infinite), then there is a feasible flow.

The proof of Theorem 1.1 and the integrity theorem establish the following fact. If the functions a, b, and c are integral valued, and if there is a feasible flow, then there is an integral feasible flow. Similar integrity statements will hold for the other feasibility theorems proved in the next two sections.

Beginning with § 4, the remainder of this chapter will require the use of such integrity statements in setting up a number of combinatorial problems as flow problems. We illustrate this approach here with the following example, suggested by Gale. Consider a round robin tournament between n teams, with each team playing every other team c times. (For instance, in pre-1961 major league baseball, $n = 8$ and $c = 22$.) Let $\alpha_i (i = 1, 2, \ldots, n)$ be the number of wins for the ith team at the conclusion of the tournament. What are necessary and sufficient conditions on a given set of non-negative integers $\alpha_1, \alpha_2, \ldots, \alpha_n$ in order that they represent a possible win record? Obvious necessary conditions are that $\sum_{i=1}^{n} \alpha_i = cn(n - 1)/2$, the total number of games played, and that $\alpha_i \leq c(n - 1)$, the total number of games played by the ith team. These conditions are of course not sufficient, since, for example, we might take $\alpha_1 = \alpha_2 = c(n - 1)$ and satisfy these conditions, yet teams 1 and 2 play each other.

To find necessary and sufficient conditions, one can proceed as follows. Select the notation so that $\alpha_1 \geq \alpha_2 \geq \ldots \geq \alpha_n \geq 0$, and define a directed network $[N; \mathscr{A}]$ by $N = \{1, 2, \ldots, n\}$, $\mathscr{A} = \{(i, j) | i < j\}$. Now, thinking of $f(i, j)$ as representing the number of wins for team i over team j, one has

$$\sum_{j>i} f(i, j) + \sum_{j<i} [c - f(j, i)] = \alpha_i$$

or

(1.7) $$f(i, N) - f(N, i) = \alpha_i - c(i - 1).$$

Conversely, any integer valued function f defined on the arcs of the network that satisfies (1.7) and

(1.8) $$0 \leq f(i, j) \leq c$$

represents a tournament in which team i wins α_i games. Defining $S \subseteq N$ and $T = \bar{S}$ by

(1.9) $$S = \{i | \alpha_i - c(i - 1) \geq 0\},$$

and corresponding supply and demand functions by

(1.10) $$a(i) = \alpha_i - c(i - 1), \qquad\qquad i \in S,$$

(1.11) $$b(i) = -\alpha_i + c(i - 1), \qquad\qquad i \in T,$$

it follows, using the integrity theorem, that the α_i represent a possible win record if and only if the constraints

(1.12) $$f(i, N) - f(N, i) = a(i), \qquad\qquad i \in S,$$

(1.13) $$f(N, i) - f(i, N) = b(i), \qquad\qquad i \in T,$$

(1.14) $$0 \leq f(i, j) \leq c,$$

are feasible. In view of the condition $\sum_{i=1}^{n} \alpha_i = cn(n - 1)/2$, which says that $a(S) = b(T)$, we can, if we like, replace (1.12) and (1.13) by inequalities (respectively $\leq$ and $\geq$) in order to obtain the supply-demand constraints

41

appearing in Theorem 1.1. Applying Theorem 1.1, it follows that the α_i represent a possible win record if, and only if, for every $X \subseteq N$,

$$(1.15) \qquad c \sum_{i \in X} (i - 1) - \sum_{i \in \bar{X}} \alpha_i \leqslant c|(X, \bar{X})|.$$

(Here $|\ |$ denotes cardinality. Note that for $\bar{X} = N$, equality actually holds in (1.15) by virtue of $\sum_{i=1}^{n} \alpha_i = cn(n - 1)/2$.)

The 2^n inequalities (1.15) can be simplified greatly. They are, in fact, equivalent to only n inequalities. To see this, first rewrite (1.15) as

$$(1.16) \qquad -c|\bar{X}| + c\left[\sum_{i \in X} i - |(X, \bar{X})| \right] \leqslant \sum_{i \in \bar{X}} \alpha_i,$$

and consider those inequalities of (1.16) for all subsets X of fixed cardinality p. The left side of (1.16) is constant for such X, being equal to $c(n - p)(n - p - 1)/2$, while the right side is minimized by taking $X = \{1, 2, \ldots, p\}$. Thus the inequalities (1.16) are equivalent to

$$(1.17) \qquad c(n - p)(n - p - 1)/2 \leqslant \sum_{i=p+1}^{n} \alpha_i, \quad p = 0, 1, \ldots, n - 1,$$

or, adding $\sum_1^n \alpha_i = cn(n - 1)/2$ to both sides, to

$$(1.18) \qquad \sum_{i=1}^{p} \alpha_i \leqslant cp(2n - p - 1)/2, \qquad p = 1, 2, \ldots, n.$$

To sum up, the necessary and sufficient conditions that $\alpha_1 \geqslant \alpha_2 \geqslant \ldots \geqslant \alpha_n \geqslant 0$ represent a win record for a round robin tournament in which each team plays c games with every other team are that the inequalities (1.18) hold, the last with equality.

2. A symmetric supply-demand theorem

Suppose that, instead of requiring the net flow out of each source to be bounded above, and the net flow into each sink to be bounded below, we extend the problem by imposing both lower and upper bounds on the net flow leaving each source and entering each sink. What are feasibility conditions for the resulting set of inequalities? One version of the theorem that will be established for this situation may be described verbally as follows:

   (a) if there is a flow that satisfies the lower bound requirements at the sources and the upper bound requirements at the sinks, and
   (b) if there is a flow that satisfies the upper bound requirements at the sources and the lower bound requirements at the sinks,

then there is a flow that meets all the requirements simultaneously.

For example, consider the network of Fig. 2.1 with all arc capacities

42

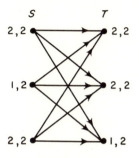

Figure 2.1

unity, the sources being the nodes on the left, the sinks on the right, with lower and upper bounds as indicated. For lower bounds at sources and upper bounds at sinks, a feasible flow is shown by the heavy arcs of Fig. 2.2, while for the reverse situation, upper bounds at sources and lower bounds at sinks, a feasible flow is shown in Fig. 2.3. Notice that the flow of

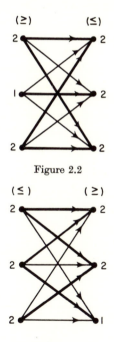

Figure 2.2

Figure 2.3

Fig. 2.2 violates the constraints of Fig. 2.3, and the flow of Fig. 2.3 violates the constraints of Fig. 2.2. According to the theorem, there is a flow meeting all constraints. One such is shown in Fig. 2.4.

43

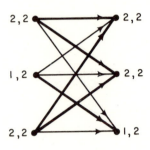

Figure 2.4

A proof of this theorem can be given along lines similar to the proof of Theorem 1.1 by transforming the given feasibility problem into an equivalent maximal flow problem in an enlarged network (using a device that will appear again in the next section). The max-flow min-cut theorem can then be applied to derive a pair of feasibility conditions, one of which is equivalent to (a) above, the other to (b).

We first describe the device to be used in transforming the feasibility problem into a maximal flow problem. Basically, what will be needed is a way of changing arbitrary lower bounds on arc flows to lower bounds that are uniformly zero. Thus suppose, for example, that in a network $[N; \mathscr{A}]$ with source s and sink t, the problem is to ascertain whether there is a flow from s to t satisfying $0 \leqslant l(x, y) \leqslant f(x, y) \leqslant c(x, y)$ for some arc (x, y). Assuming that all source arcs are directed from s and all sink arcs into t, the problem may be pictured schematically as in Fig. 2.5. (The possibility

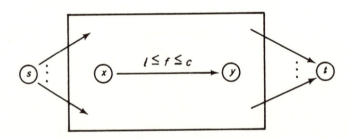

Figure 2.5

$s = x$ or $y = t$ is not excluded.) Enlarge the network as suggested in Fig. 2.6 by adding two nodes u, w, the arcs (u, y), (x, w), each having capacity $l(x, y)$, the arc (t, s) having infinite capacity, and let (x, y) have the new capacity $c(x, y) - l(x, y)$. Then a feasible flow f from s to t of value

44

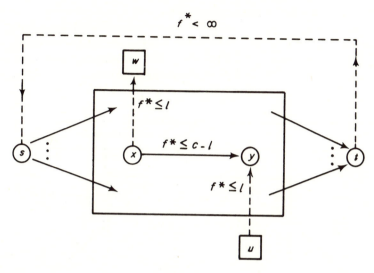

Figure 2.6

v in the original network generates a flow f^* from u to w of value $l(x, y)$ by defining

$$f^*(t, s) = v,$$
$$f^*(u, y) = f^*(x, w) = l(x, y),$$
$$f^*(x, y) = f(x, y) - l(x, y),$$
$$f^* = f, \qquad\qquad\qquad \text{otherwise,}$$

and conversely. Thus a feasible flow exists if and only if the value of a maximal flow in the new network is $l(x, y)$.

Another way to interpret this device is to think of a supply $l(x, y)$ at y and a demand $l(x, y)$ at x, thereby eliminating the nodes u, w and their arcs.

Returning now to the original problem of finding feasibility conditions in $[N; \mathscr{A}]$ for the constraints

$$a(x) \leqslant f(x, N) - f(N, x) \leqslant a'(x), \qquad x \in S,$$
$$f(x, N) - f(N, x) = 0, \qquad x \in R,$$
$$b(x) \leqslant f(N, x) - f(x, N) \leqslant b'(x), \qquad x \in T,$$
$$0 \leqslant f(x, y) \leqslant c(x, y), \qquad (x, y) \in \mathscr{A},$$

where a, a', b, b' are given functions satisfying

$$0 \leqslant a(x) \leqslant a'(x), \qquad x \in S,$$
$$0 \leqslant b(x) \leqslant b'(x), \qquad x \in T,$$

first extend the network by adjoining new nodes s, t and the arcs (s, S), (T, t), where (s, x), $x \in S$, has lower and upper bounds $a(x)$, $a'(x)$; and

45

(x, t), $x \in T$, has lower and upper bounds $b(x)$, $b'(x)$. A further extension can then be made using the device discussed above for getting rid of non-zero lower bounds on arc flows. The result, pictured schematically in Fig. 2.7, is a network $[N^*; \mathscr{A}^*]$, where N^* consists of N plus four new nodes s, t, u, w and $\mathscr{A}^*$ consists of $\mathscr{A}$ and the additional arcs (s, S), (u, S), (T, t), (T, w), (u, t), (s, w), (t, s).

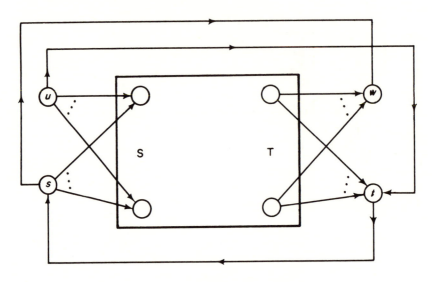

Figure 2.7

The capacity function c defined on $\mathscr{A}$ is extended to $\mathscr{A}^*$ by

$$c(s, x) = a'(x) - a(x), \qquad x \in S,$$
$$c(u, x) = a(x), \qquad x \in S,$$
$$c(x, t) = b'(x) - b(x), \qquad x \in T,$$
$$c(x, w) = b(x), \qquad x \in T,$$
$$c(u, t) = b(T),$$
$$c(s, w) = a(S),$$
$$c(t, s) = \infty.$$

We assert that a feasible flow exists in $[N; \mathscr{A}]$ if and only if the value of a maximal flow from u to w in $[N^*; \mathscr{A}^*]$ is $a(S) + b(T)$. Suppose first that f is feasible in $[N; \mathscr{A}]$. Extend f to f^*, defined on $\mathscr{A}^*$, as follows:

$$f^*(s, x) = f(x, N) - f(N, x) - a(x), \qquad x \in S,$$
$$f^*(u, x) = a(x), \qquad x \in S,$$
$$f^*(x, t) = f(N, x) - f(x, N) - b(x), \qquad x \in T,$$
$$f^*(x, w) = b(x), \qquad x \in T,$$

$$f^*(u, t) = b(T),$$
$$f^*(s, w) = a(S),$$
$$f^*(t, s) = f(S, N) - f(N, S),$$
$$f^*(x, y) = f(x, y), \qquad\qquad (x, y) \in \mathscr{A}.$$

It is a routine matter to check that f^* is a flow from u to w in $[N^*; \mathscr{A}^*]$. Clearly, f^* has value $a(S) + b(T)$.

Conversely, let f^* be a flow from u to w in $[N^*; \mathscr{A}^*]$ of value $a(S) + b(T)$. Then

$$f^*(u, x) = a(x), \qquad\qquad x \in S,$$
$$f^*(x, w) = b(x), \qquad\qquad x \in T.$$

Let f be f^* restricted to $\mathscr{A}$. Then f is a flow from S to T in $[N; \mathscr{A}]$, and it remains only to show that f is feasible. Suppose $x \in S$. Then

$$f^*(u, x) + f^*(s, x) = f(x, N) - f(N, x),$$

or

$$a(x) + f^*(s, x) = f(x, N) - f(N, x);$$

and, since $0 \leqslant f^*(s, x) \leqslant a'(x) - a(x)$, we get

$$a(x) \leqslant f(x, N) - f(N, x) \leqslant a'(x).$$

The inequalities

$$b(x) \leqslant f(N, x) - f(x, N) \leqslant b'(x), \qquad\qquad x \in T,$$

are similarly proved. This completes the proof of the assertion.

We may, therefore, in searching for feasibility criteria, re-phrase the question as follows. Under what conditions does there exist a flow f^* from u to w in $[N^*; \mathscr{A}^*]$ having value $a(S) + b(T)$, that is, saturating all source and sink arcs?

The max-flow min-cut theorem can now be used to provide an answer to this question by insisting that the capacities of all cuts separating u and w be at least $a(S) + b(T)$. Thus, let $(X^*, \overline{X}^*)$ be a cut in $[N^*; \mathscr{A}^*]$ and consider cases.

Case 1. $s \in X^*$, $t \in \overline{X}^*$. Partition X^*, $\overline{X}^*$ as follows: $X^* = u \cup s \cup X$, $\overline{X}^* = w \cup t \cup \overline{X}$, so that $\overline{X}$ is the complement of X in N. Then

$$c(X^*, \overline{X}^*) = c(u, t) + c(u, \overline{X}) + c(s, w) + c(S, \overline{X}) + c(X, w) + c(X, t)$$
$$+ c(X, \overline{X})$$
$$= b(T) + a(S \cap \overline{X}) + a(S) + a'(S \cap \overline{X}) - a(S \cap \overline{X})$$
$$+ b(T \cap X) + b'(T \cap X) - b(T \cap X) + c(X, \overline{X}).$$

Hence in this case, we always have $c(X^*, \overline{X}^*) \geqslant a(S) + b(T)$.

Case 2. $s \in \overline{X}^*$, $t \in X^*$. Then $c(X^*, \overline{X}^*)$ is infinite, and again no condition is obtained.

Case 3. $s \in X^*, t \in X^*$. Letting $X^* = s \cup t \cup u \cup X$, $\overline{X}^* = w \cup \overline{X}$, we have

$$c(X^*, \overline{X}^*) = c(s, w) + c(s, \overline{X}) + c(u, \overline{X}) + c(X, w) + c(X, \overline{X})$$
$$= a(S) + a'(S \cap \overline{X}) - a(S \cap \overline{X}) + a(S \cap \overline{X}) + b(T \cap X)$$
$$+ c(X, \overline{X}).$$

Thus $c(X^*, \overline{X}^*) \geqslant a(S) + b(T)$ if and only if

$$c(X, \overline{X}) \geqslant b(T \cap \overline{X}) - a'(S \cap \overline{X}).$$

Case 4. $s \in \overline{X}^*, t \in \overline{X}^*$. Let $X^* = u \cup X$, $\overline{X}^* = s \cup t \cup w \cup \overline{X}$. Then

$$c(X^*, \overline{X}^*) = c(u, t) + c(u, \overline{X}) + c(X, t) + c(X, w) + c(X, \overline{X})$$
$$= b(T) + a(S \cap \overline{X}) + b'(T \cap X) - b(T \cap X) + b(T \cap X)$$
$$+ c(X, \overline{X}),$$

and we obtain the condition

$$c(X, \overline{X}) \geqslant a(S \cap X) - b'(T \cap X).$$

We may therefore state the following result [8].

THEOREM 2.1. *The constraints*

(2.1) $$a(x) \leqslant f(x, N) - f(N, x) \leqslant a'(x), \qquad x \in S,$$

(2.2) $$f(x, N) - f(N, x) = 0, \qquad x \in R,$$

(2.3) $$b(x) \leqslant f(N, x) - f(x, N) \leqslant b'(x), \qquad x \in T,$$

(2.4) $$0 \leqslant f(x, y) \leqslant c(x, y), \qquad (x, y) \in \mathscr{A},$$

(where $0 \leqslant a(x) \leqslant a'(x)$ for $x \in S$ and $0 \leqslant b(x) \leqslant b'(x)$ for $x \in T$) are feasible if and only if

(2.5) $$c(X, \overline{X}) \geqslant b(T \cap \overline{X}) - a'(S \cap \overline{X}),$$

(2.6) $$c(X, \overline{X}) \geqslant a(S \cap X) - b'(T \cap X)$$

hold for all $X \subseteq N$.

Notice that (2.5) is precisely condition (1.5) for the supply-demand case (Theorem 1.1); that is, if $a(x) = 0$ for $x \in S$ and $b'(x) = \infty$ for $x \in T$, then Theorem 2.1 reduces to Theorem 1.1. Condition (2.6) may be interpreted as follows. If we interchange sources and sinks in $[N; \mathscr{A}]$, reverse all arc orientations, and think of a as the demand function at the set S of sinks, b' as the supply function at the set T of sources, then (2.6) is a necessary and sufficient condition for feasibility of the supplies and demands in the reversed network. Thus Theorem 2.1 may be restated as follows.

COROLLARY 2.2.   *The constraints (2.1)–(2.4) are feasible if and only if each of the constraint sets*

(2.7)
$$a(x) \leqslant f(x, N) - f(N, x), \qquad x \in S,$$
$$f(x, N) - f(N, x) = 0, \qquad x \in R,$$
$$f(N, x) - f(x, N) \leqslant b'(x), \qquad x \in T,$$
$$0 \leqslant f(x, y) \leqslant c(x, y), \qquad (x, y) \in \mathscr{A},$$

(2.8)
$$f(x, N) - f(N, x) \leqslant a'(x), \qquad x \in S,$$
$$f(x, N) - f(N, x) = 0, \qquad x \in R,$$
$$b(x) \leqslant f(N, x) - f(x, N), \qquad x \in T,$$
$$0 \leqslant f(x, y) \leqslant c(x, y), \qquad (x, y) \in \mathscr{A},$$

is feasible.

Corollary 2.2 is the formulation described verbally at the beginning of this section.

When the network is suitably specialized, Theorem 2.1 (or its corollary) provides criteria for the existence of a non-negative matrix whose row and column sums lie between designated limits, or, more generally, for the existence of a matrix with these properties and the further property that the elements of the matrix are bounded above by specified numbers. We state the criteria provided by Corollary 2.2 explicitly.

COROLLARY 2.3.   *Let* $0 \leqslant a_i \leqslant a_i'$, $i = 1, \ldots, m$, $0 \leqslant b_j \leqslant b_j'$, $j = 1$, $\ldots, n$, *and* $c_{ij} \geqslant 0$ *be given constants. If there are matrices* (f_{ij}^1), (f_{ij}^2) *satisfying*

(2.9)
$$a_i \leqslant \sum_j f_{ij}^1, \qquad \sum_i f_{ij}^1 \leqslant b_j', \qquad 0 \leqslant f_{ij}^1 \leqslant c_{ij},$$

(2.10)
$$\sum_j f_{ij}^2 \leqslant a_i', \qquad b_j \leqslant \sum_i f_{ij}^2, \qquad 0 \leqslant f_{ij}^2 \leqslant c_{ij},$$

then there is a matrix (f_{ij}) *satisfying*

(2.11) $$a_i \leqslant \sum_j f_{ij} \leqslant a_i', \qquad b_j \leqslant \sum_i f_{ij} \leqslant b_j', \qquad 0 \leqslant f_{ij} \leqslant c_{ij}.$$

To prove Corollary 2.3, take $[N; \mathscr{A}]$ to be the network consisting of nodes $x_i (i = 1, \ldots, m)$, y_j $(j = 1, \ldots, n)$, and arcs (x_i, y_j) of capacity c_{ij}. Let $S = \{x_1, \ldots, x_m)$, $T = \{y_1, \ldots, y_n\}$, so that R is vacuous. Associate with each source x_i the bounds a_i, a_i' and with each sink y_j the bounds b_j, b_j'. Then a flow from S to T is a matrix (f_{ij}) satisfying $0 \leqslant f_{ij} \leqslant c_{ij}$; a feasible flow satisfies, in addition, the first two inequalities of (2.11). Thus Corollary 2.3 is an immediate consequence of Corollary 2.2.

The particular kind of network involved in the proof of Corollary 2.3, namely one in which the nodes N are divided into two subsets with all arcs of $\mathscr{A}$ leading from nodes of one subset to those of the other (we do not, however, insist that all such arcs be present in $\mathscr{A}$), will crop up frequently enough to justify a special name. One name that has been used by writers in graph theory is "bipartite"; henceforth we shall use this terminology also.

3. Circulation theorem

The final feasibility theorem that we shall discuss before proceeding to combinatorial applications is due to Hoffman [17]. This theorem is concerned with the existence of circulations; that is, flows that are source and sink free, that satisfy prescribed lower and upper bounds on arcs.

The method used in the last section can be applied to this problem as well.

Let the given network be $[N; \mathscr{A}]$ and suppose that l and c are the lower and upper bound functions defined on $\mathscr{A}$, where $0 \leqslant l \leqslant c$. A *feasible circulation* in $[N; \mathscr{A}]$ is a function f on $\mathscr{A}$ satisfying

$$f(x, N) - f(N, x) = 0, \qquad x \in N,$$
$$l(x, y) \leqslant f(x, y) \leqslant c(x, y), \qquad (x, y) \in \mathscr{A}.$$

Extend $[N; \mathscr{A}]$ to $[N^*; \mathscr{A}^*]$ by the adjunction of two nodes s, t and the sets of arcs (s, N) and (N, t). The capacity function defined on $\mathscr{A}^*$ is

$$c^*(x, y) = c(x, y) - l(x, y), \qquad (x, y) \in \mathscr{A},$$
$$c^*(s, x) = l(N, x), \qquad x \in N,$$
$$c^*(x, t) = l(x, N), \qquad x \in N.$$

It is easy to verify that a feasible circulation f in $[N; \mathscr{A}]$ generates a flow f^* from s to t in $[N^*; \mathscr{A}^*]$ via the rule:

$$f^*(x, y) = f(x, y) - l(x, y), \qquad (x, y) \in \mathscr{A},$$
$$f^*(s, x) = l(N, x), \qquad x \in N,$$
$$f^*(x, t) = l(x, N), \qquad x \in N.$$

Thus the question becomes: when is there a flow from s to t in $[N^*; \mathscr{A}^*]$ that has value $l(N, N)$?

The procedure is now familiar. The necessary and sufficient condition for the existence of a flow from s to t of value $l(N, N)$ is that all cut capacities exceed $l(N, N)$. Let $(X^*, \overline{X}^*)$ be a cut separating s and t in $[N^*; \mathscr{A}^*]$, and define $X \subseteq N$ together with its complement $\overline{X}$ in N by

$$X = X^* - s, \overline{X} = \overline{X}^* - t.$$

Then

$$c^*(X^*, \overline{X}^*) = c^*(X \cup s, \overline{X} \cup t)$$
$$= c^*(X, \overline{X}) + c^*(s, \overline{X}) + c^*(X, t)$$
$$= c(X, \overline{X}) - l(X, \overline{X}) + l(N, \overline{X}) + l(X, N)$$
$$= c(X, \overline{X}) + l(\overline{X}, \overline{X}) + l(X, N),$$

and consequently $c^*(X^*, \overline{X}^*) \geqslant l(N, N)$ if and only if

$$c(X, \overline{X}) \geqslant l(\overline{X}, X).$$

Thus we have established

THEOREM 3.1. *A necessary and sufficient condition for the constraints*

(3.1) $f(x, N) - f(N, x) = 0,$ $x \in N,$

(3.2) $l(x, y) \leqslant f(x, y) \leqslant c(x, y),$ $(x, y) \in \mathscr{A},$

to be feasible, where $0 \leqslant l(x, y) \leqslant c(x, y),$ *is that*

(3.3) $c(X, \overline{X}) \geqslant l(\overline{X}, X)$

hold for all $X \subseteq N.$

As in the other feasibility theorems, the necessity of the condition is intuitively clear (and easily proved directly), since (3.3) simply asserts that there must be sufficient escape capacity from the set X to take care of the flow forced into X by the function l. If the condition is satisfied for all subsets of nodes, the existence of a feasible circulation is assured.

The reason why the necessity of such feasibility conditions is always the easier half of the theorem is that necessity corresponds to the weak half of the max-flow min-cut theorem, that is, flow values are bounded above by cut capacities.

There is one fundamental difference between the supply-demand theorems and the circulation theorem, however, that lies in the distinction between flows in undirected or mixed networks and flows in directed networks. If we interpret flows in undirected or mixed networks to mean that flow in an arc is unidirectional but no direction may be specified (as in I.10), then Theorems 1.1 and 2.1 remain valid. But Theorem 3.1 gives no information about the problem of determining conditions under which a feasible circulation exists in an undirected network subject to lower and upper bounds on arc flows. (This problem is, so far as we know, unsolved.) Generally speaking, if non-zero lower bounds are imposed on undirected arc flows, then replacing an undirected arc by a pair of oppositely directed arcs and cancelling flows in opposite directions, is not a valid operation. Thus, there is a real distinction between directed and undirected problems in this case.

The circulation theorem can be used to answer the question raised in I.9 concerning the existence of flows from s to t in a directed network subject to lower and upper bounds on arc flows. By adding the arcs (s, t) and (t, s) to the network with infinite capacity (allowing multiple arcs if necessary), the relevant condition becomes (3.3) for all $X \subseteq N$ such that either both s and t belong to X or neither does. To construct such flows, one can of course solve the equivalent maximal flow problem used in deriving the existence conditions.

51

Hoffman has also stated an extension of the circulation theorem covering the situation in which the net flow into node x lies between stipulated bounds [17]. The result here, which may be proved either from the circulation theorem or its variant described above, is:

THEOREM 3.2. *The constraints*

(3.4) $$a(x) \leqslant f(N, x) - f(x, N) \leqslant a'(x),$$

(3.5) $$l(x, y) \leqslant f(x, y) \leqslant c(x, y),$$

(where $a(x) \leqslant a'(x)$, $0 \leqslant l(x, y) \leqslant c(x, y)$) are feasible if and only if

(3.6) $$c(X, \bar{X}) \geqslant l(\bar{X}, X) + \max [a(\bar{X}), - a'(X)]$$

holds for all $X \subseteq N$.

A proof of Theorem 3.2 can be given by adjoining a source s and sink t to N, the sets of arcs (s, N) and (N, t) to $\mathscr{A}$, and extending the functions l and c to arcs of the new network by defining

$$l(s, x) = \max (0, -a'(x)),$$
$$l(x, t) = \max (0, a(x)),$$
$$c(s, x) = \max (0, -a(x)),$$
$$c(x, t) = \max (0, a'(x)).$$

Applying the feasibility conditions stated in the preceding paragraph to the new network yields the pair of conditions embodied in (3.6).

In case $a(x) = a'(x) = 0$, or indeed, if $a(x) = 0$, then (3.6) reduces to (3.3).

Although the proof of Theorem 3.1 that has been presented in this section used the max-flow min-cut theorem, an alternate, direct proof can be given along lines similar to the proof of the latter. The direct construction for a feasible circulation described below provides such a proof in case the lower bound function l and capacity function c are rational valued. The basic routine in this construction is again a labeling process.

We assume for the construction that l and c are integral valued.

Construction of a feasible circulation. Start with any integral valued f that satisfies the conservation equations at all nodes. For example, $f = 0$ will do. Next locate an arc (s, t) for which one of the bound conditions (3.2) is violated, and go on to the appropriate case below.

Case 1. $f(s, t) > c(s, t)$. Start a labeling process at node s, trying to reach node t, assigning labels to nodes as follows. First label s with $[t^-, \varepsilon(s) = f(s, t) - c(s, t)]$. In general, select any labeled, unscanned node x, and assign further labels to (unlabeled) nodes y using the rules:

(a) if (x, y) is an arc with $f(x, y) < c(x, y)$, assign y the label $[x^+, \varepsilon(y) = \min (\varepsilon(x), c(x, y) - f(x, y))]$;

(b) if (y, x) is an arc with $f(y, x) > l(y, x)$, assign y the label $[x^-_\cdot, \varepsilon(y) = \min (\varepsilon(x), f(y, x) - l(y, x))]$.

Continue labeling until either node t is labeled (breakthrough), or until no more labels can be assigned and node t is unlabeled (non-breakthrough). In the former case, change the flow by adding and subtracting $\varepsilon(t)$ while back-tracking from t to s according to first members of the labels; having reached s, also subtract $\varepsilon(t)$ from $f(s, t)$. If non-breakthrough occurs, terminate. (There is no feasible circulation.)

Case 2. $f(s, t) < l(s, t)$. Start labeling at t, trying to reach s, first assigning t the label $[s^+, \varepsilon(t) = l(s, t) - f(s, t)]$. The labeling rules are the same as *Case* 1. If breakthrough occurs, so that a path from t to s has been found, change the flow by adding $\varepsilon(s)$ to the flow in forward arcs of this path, subtracting $\varepsilon(s)$ from the flow in reverse arcs, and adding $\varepsilon(s)$ to $f(s, t)$. If non-breakthrough, terminate. (There is no feasible circulation.)

Following a flow change in either case, locate another arc flow that violates its bounds, and re-label.

This algorithm either constructs a feasible circulation in finitely many steps, or proves there is no feasible circulation. First of all, note that if breakthrough occurs in either case, a cycle has been found that includes the arc (s, t). (For in *Case* 1, t cannot be labeled from s via the arc (s, t), and similarly in *Case* 2.) Then the flow change made on arcs of this cycle again yields an f' satisfying the conservation equations. Moreover, the new arc flow $f'(s, t)$ comes at least one unit closer to feasibility, and if $l(x, y) \leqslant f(x, y) \leqslant c(x, y)$ for any other (x, y), then also $l(x, y) \leqslant f'(x, y) \leqslant c(x, y)$. It follows that, after finitely many steps, either a feasible circulation is constructed, or non-breakthrough occurs.

Suppose that non-breakthrough occurs, say in *Case* 1, and let X and $\overline{X}$ be the labeled and unlabeled sets of nodes. Then $s \in X$, $t \in \overline{X}$. It follows from the labeling rules that $f(x, \overline{x}) \geqslant c(x, \overline{x})$ for all arcs in $(X, \overline{X})$, and $f(\overline{x}, x) \leqslant l(\overline{x}, x)$ for all arcs of $(\overline{X}, X)$. Also, for at least one arc of $(X, \overline{X})$, namely (s, t), we have strict inequality $f(s, t) > c(s, t)$. Thus, since f satisfies the conservation equations at all nodes,

$$0 = f(X, \overline{X}) - f(\overline{X}, X) > c(X, \overline{X}) - l(\overline{X}, X),$$

violating condition (3.3). Hence there is no feasible circulation.

An exactly similar proof holds for *Case* 2.

A variant of this construction for a feasible circulation will play a role in one of the algorithms of Chapter III.

4. The König-Egerváry and Menger graph theorems

We shall customarily use the word "graph" or "linear graph" when our intent is to focus attention on purely combinatorial results, and use "network" when the primary concern is with flows.

There are two well-known theorems in linear graph theory that are intimately related to the max-flow min-cut theorem for network flows.

They may, in fact, be regarded as combinatorial prototypes of the latter. The first of these theorems (Theorem 4.1 below), due to König and Egerváry, appears as a lemma in the proof of Menger's theorem given in [21]; it deals with bipartite graphs. Menger's theorem (Theorem 4.2 below) is the generalization that results for arbitrary graphs.

We use the notation $[N ; \mathscr{A}] = [S, T ; \mathscr{A}]$ for bipartite graphs in which arcs lead from S to T.

THEOREM 4.1. *Let $G = [S, T ; \mathscr{A}]$ be a bipartite graph. The maximal number of arcs of G that are pairwise node disjoint is equal to the minimal number of nodes in an S, T disconnecting set of nodes.*†

Here an S, T disconnecting set of nodes is a set of nodes that blocks all chains from S to T.

To prove this theorem using flows, one can proceed as follows. Adjoin nodes s, t and the sets of arcs (s, S), (T, t) to the network. For the resulting network $[N^* ; \mathscr{A}^*]$ define a capacity function by

$$
\begin{aligned}
c(s, x) &= 1, & x &\in S, \\
c(x, t) &= 1, & x &\in T, \\
c(x, y) &= \infty, & (x, y) &\in \mathscr{A}.
\end{aligned}
$$

Let f be an integral maximal flow from s to t and let $(X, \overline{X})$ be a minimal cut separating s and t. (Note that $(X, \overline{X})$ can contain no arcs of $\mathscr{A}$.) The arcs of the set $\mathscr{I} = \{(x, y) \in \mathscr{A} | f(x, y) = 1\}$ are pairwise node disjoint, and the nodes of the S, T disconnecting set $D = (S \cap \overline{X}) \cup (T \cap X)$ are in one-one correspondence with the arcs of the minimal cut $(X, \overline{X})$. It follows from the max-flow min-cut theorem that if f has value v, then v is the number of elements in $\mathscr{I}$, and also in D. Hence, since the maximal number of pairwise node disjoint arcs of G is clearly less than or equal to the minimal number of nodes in an S, T disconnecting set of nodes, the proof of Theorem 4.1 is complete.

Another statement of the König-Egerváry theorem is sometimes given in terms of m by n arrays that contain two kinds of cells, admissible and inadmissible, say. Suppose we refer to the rows and columns of the array by the common term "lines." A set of lines *covers* the admissible cells of the array if each admissible cell belongs to some line of the set. A set of admissible cells is *independent* if no two cells of the set lie in the same line. By constructing from the array the bipartite graph G composed of nodes

$$
S = \{x_1, \ldots, x_m\}, \qquad T = \{y_1, \ldots, y_n\},
$$

† Many combinatorial proofs of this theorem are known. Of these, perhaps the one closest in spirit to the use of flows is that of [21], in which the notion of an alternating path substitutes for that of a flow augmenting path. One other proof we wish to call the reader's attention to is given in [5]. This proof, and the use of it made by Kuhn [22] in devising an algorithm for the optimal assignment problem, played an important role in the development of the algorithms to be presented later for minimal cost transportation problems.

and arcs (x_i, y_j) corresponding to admissible cells, one sees that the notion of "independent set of admissible cells" (respectively, "covering set of lines") corresponds to "pairwise node disjoint arcs" (respectively, "S, T disconnecting set of nodes"), and hence Theorem 4.1 becomes: the maximal number of independent admissible cells is equal to the minimal number of lines that cover all admissible cells.

THEOREM 4.2. *Let S and T be two disjoint subsets of the nodes of the graph $G = [N; \mathscr{A}]$. The maximal number of pairwise node disjoint chains from S to T is equal to the minimal number of nodes in an S, T disconnecting set of nodes.*

Again this theorem follows from the max-flow min-cut theorem and integrity theorem by adjoining a source s and sink t, together with source arcs (s, S) and sink arcs (T, t), and imposing unit capacity on all old nodes, infinite capacity on arcs. A chain decomposition of an integral maximal flow from s to t provides a maximal set of pairwise node disjoint chains.

The graph G in Theorem 4.2 may be directed, undirected, or mixed without affecting the theorem statement. It is also clear that a similar theorem holds for chains from S to T that are pairwise arc disjoint and sets of arcs that block all chains from S to T, since we may place unit capacity on arcs, infinite capacity on nodes.

The max-flow min-cut theorem is obviously a generalization of Theorem 4.2. On the other hand, a proof of the max-flow min-cut theorem that uses Theorem 4.2 as the principal tool has been given by Robacker [28].

5. Construction of a maximal independent set of admissible cells

The labeling process for constructing maximal flows can of course be used to produce a maximal independent set of admissible cells and a minimal covering set of lines for the array interpretation of Theorem 4.1. It is worth while describing this computation in detail, since some simplification is possible because of the special nature of the associated flow problem. The algorithm that results is similar to the construction that may be considered implicit in König's proof of Theorem 4.1, and also similar in spirit, although not in detail, to Kuhn's method for solving this problem [22].

Let $i = 1, \ldots, m$ index the rows of the array, $j = 1, \ldots, n$ the columns. The maximal flow problem becomes that of placing as many 1's as possible in admissible cells, with the proviso that at most one 1 can be placed in any line. Initiate the process with any feasible placement of 1's, e.g., scan the first row and place a 1 in the first admissible cell, then delete the lines containing this 1 and repeat the procedure in the reduced array.

After obtaining a feasible placement of 1's, begin by labeling (with dashes, say) all rows that contain no 1's. Then select a labeled row, say the ith row, and scan it for admissible cells, labeling the (unlabeled) columns corresponding to such cells by the number of the row being scanned, here i. Repeat until all labeled rows have been scanned (never labeling a column that has already received a label). Now select any labeled column, say column j, and scan it for a 1; if such is found, label the row in which the 1 lies with the number of the column being scanned, here j. Again select an unscanned, labeled column and repeat the procedure. After scanning all labeled columns, revert to row scanning by selecting a labeled, unscanned row. The process continues in this fashion, alternating between row and column scanning until either:

(a) a column is labeled that contains no 1, in which case an improved placement of 1's can be found from the labels (breakthrough);

(b) no more labels are possible and breakthrough has not occurred, in which case the present placement of 1's is maximal (non-breakthrough).

In case (a) the total number of 1's in the array can be increased (by one) as follows. In the column containing no 1 that has just been labeled, place a 1 in the position designated by its label; then proceed, in the row in which this 1 lies, to the position indicated by its label and remove the 1 there; then go, in the column just reached, to the position indicated by its label, and place a 1, and so on. Eventually one of the initially labeled rows (those marked by dashes) will be reached, at which point the replacement stops, and the total number of 1's in the array has been increased by one. The labeling process is then repeated with the new placement of 1's.

In case (b), a minimal covering set of lines consists of the unlabeled rows and labeled columns.

EXAMPLE. In the array of Fig. 5.1, admissible cells are blank and inadmissible cells are crossed out. The 1's shown constitute an initial placement using the suggested starting procedure. Rows 8 and 9 have no 1's; we need to "break through" to either column 5 or 9 to get an improvement. Scanning row 8 labels columns 2, 4, 6; row 9 produces the additional labels on columns 1, 8. From column 1 we label row 2; from column 2, row 1; from column 4, row 3; from column 6, row 5; and from column 8, row 6. Switching back to row scanning, we get only the additional label 2 on column 3. Then row 4 receives the label 3, following which breakthrough into either column 5 or 9 occurs; here we have labeled column 5. The arrows in Fig. 5.1 indicate the resulting sequence of changes in the placement of 1's.

After making the indicated changes and relabeling, case (b) occurs and a minimal cover is found to consist of rows 2, 4, 7, 9 and columns 2, 4, 6, 8.

A practical instance of this kind of problem might occur, for example, in attempting to fill jobs with qualified personnel. Thus if man i is qualified

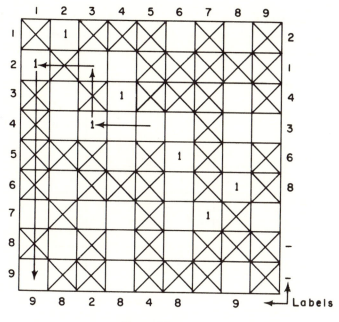

Figure 5.1

for job j, cell ij is classified "admissible," otherwise "inadmissible." An assignment of men to jobs that maximizes the number of men assigned to jobs they are qualified for can then be found by the procedure we have outlined. As we shall see later, this problem is a special case of the optimal assignment problem; a solution to the latter can, however, be obtained by solving a sequence of such special assignment problems.

6. A bottleneck assignment problem

The computation of the preceding section can also be applied repeatedly to solve the following bottleneck problem. Suppose there are n men and n jobs, that man i in job j has an "efficiency" a_{ij}, and that it is desired to find an assignment $i \to P(i)$ of men to jobs that maximizes the least $a_{i,\ P(i)}$; that is, we want to construct a permutation P^* that achieves $\max_P \min_i a_{i,\ P(i)}$. For example, the jobs might be those on an assembly line, and a_{ij} might represent the number of units per hour that man i can process if assigned to job j. Then for a given assignment P, the rate of the assembly line is measured by the bottleneck $\min_i a_{i,\ P(i)}$, and thus we wish to maximize this over all permutations P.

Gross has pointed out a simple procedure for solving this problem [12]. Briefly, it is this. Begin by selecting an arbitrary permutation P. Then, in

57

the array (a_{ij}), call a cell admissible or inadmissible according as a_{ij} $> \min_i a_{i,P(i)}$ or $a_{ij} \leqslant \min_i a_{i,P(i)}$. (Clearly P can be improved if and only if n independent admissible cells can now be found in the array.) Apply the algorithm of § 5 to construct a maximal set of k admissible cells. If $k < n$, P is optimal; if $k = n$, repeat the procedure with the new permutation thus defined.

EXAMPLE. Suppose the array (a_{ij}) is that of Fig. 6.1 and we initiate

1	3✓	2	6	0	1
4✓	2	3	8	3	1
8	1	1	5✓	0	9
3	5	4✓	8	8	3
2	6	9	5	2	4✓
3	2	3	6	7✓	1

Figure 6.1

the computation with the permutation indicated by the checks ($\checkmark$). The resulting admissible cells are indicated by circles in Fig. 6.2. We may then

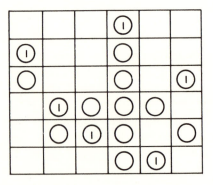

Figure 6.2

start with the partial assignment obtained by retaining as many checks in admissible cells as possible (here five) and apply the labeling procedure to construct the assignment indicated by 1's in Fig. 6.2. In the new array of admissible cells thus defined (Fig. 6.3), there are at most five independent admissible cells, and hence the assignment of Fig. 6.2 is optimal.

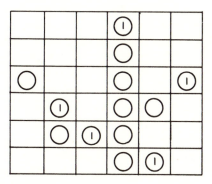

Figure 6.3

7. Unicursal graphs

A mixed linear graph $G = [N; \mathscr{A}]$ will be called *unicursal* if there exists a closed route in G that contains each arc of G once and only once, and each node of G at least once. Here the phrase "closed route" means a sequence of nodes and arcs that has the form

$$(7.1) \qquad x_1, (x_1, x_2), x_2, (x_2, x_3), \ldots, (x_{n-1}, x_1), x_1.$$

Thus a closed route differs from a directed cycle in that the nodes of (7.1) need not be distinct; but any directed arc encountered in traversing a closed route will be traversed with its direction, as for a directed cycle. To avoid special statements, we also stipulate that a single node is a closed route. Hence the graph consisting of one node and no arcs is unicursal.

The study of graphs had its origin in unicursal problems; in particular, Euler's celebrated problem of the "bridges of Königsberg." Necessary and sufficient conditions that a given graph G be unicursal are well known in case G is either directed or undirected. In this section we shall use flows to derive such conditions for mixed graphs. These conditions contain as special cases those for the directed or undirected case.

To state Theorem 7.1 below, we require one other elementary notion about graphs that has not been previously elaborated, that of connectedness. For our purposes here, we may say that two distinct nodes x and y of a graph G are *connected* if there is a path in G from x to y (and hence from y to x). This defines an equivalence relation on the nodes of G, and therefore partitions N into connected classes $N_1, N_2, \ldots, N_p$ which have the properties:

(1) every two nodes of the same class are connected,
(2) there is no arc in G joining nodes of distinct classes.

59

If $p = 1$, the graph G is *connected*; that is, every two nodes of G are connected.† Thus every graph G splits up into connected subgraphs $H_1, H_2, \ldots, H_p$. Here H_i is the subgraph of G consisting of all nodes of the ith equivalence class N_i, together with all arcs of G that join nodes of this class.

It is clear that a necessary condition for unicursality is that the graph be connected. Another obvious necessary condition is that every node of the graph be incident with an even number of arcs, since a closed route leaves a node as often as it enters it. For undirected graphs, these two necessary conditions are also sufficient. In order to obtain necessary and sufficient conditions for unicursality for directed graphs, however, the latter condition must be replaced by the condition that the number of inwardly directed arcs is equal to the number of outwardly directed arcs at each node.

THEOREM 7.1. *The mixed graph $G = [N; \mathscr{A}]$ is unicursal if and only if*

(a) *G is connected;*

(b) *every node of G is incident with an even number of arcs;*

(c) *for every $X \subseteq N$, the difference between the number of directed arcs from X to $\overline{X}$ and the number of directed arcs from $\overline{X}$ to X is less than or equal to the number of undirected arcs joining X and $\overline{X}$.*

Notice that in case G is directed or undirected, the conditions of the theorem reduce to those stated above for these cases. We now prove Theorem 7.1. Again necessity gives no difficulty. To prove sufficiency, one can proceed by establishing a circulation in G, then directing some of the originally undirected arcs according to this circulation.

In particular, first replace each undirected arc of G by a pair of oppositely directed arcs, obtaining a directed graph $G_1 = [N; \mathscr{A}_1]$. Define lower bounds and capacities for arcs (x, y) in $\mathscr{A}_1$ by

$$(7.2) \qquad\qquad c(x, y) = 1, \qquad\qquad (x, y) \in \mathscr{A}_1,$$

$$(7.3) \qquad\qquad l(x, y) = \begin{cases} 1, & \text{if } (x, y) \text{ is a directed arc of } \mathscr{A}, \\ 0, & \text{otherwise.} \end{cases}$$

Then hypothesis (c) of the theorem is equivalent to (3.3) of the circulation theorem, and hence there is a feasible circulation f in G_1. This circulation may further be assumed integral, hence $f(x, y) = 0$ or 1. Now orient some of the undirected arcs of G as follows. If (x, y) is an undirected arc of G and if $f(x, y) = 1$, $f(y, x) = 0$, direct the arc from x to y. This yields a mixed graph $G_2 = [N; \mathscr{A}_2]$, which has properties (a), (b) of the theorem, and also

---

† G might better be said to be "weakly" connected, reserving the notion of "strong" connectedness for a graph in which there is a chain from any node to another. Actually, it is immaterial whether (a) of Theorem 7.1 is stated in terms of strong or weak connectedness.

(c′) *the number of inwardly directed arcs equals the number of outwardly directed arcs at each node.*

It now suffices to establish that G_2 is unicursal. To do this, one can proceed by induction on the number of arcs of such a graph. If G_2 has no arcs and is connected, it consists of a single node, and is unicursal. Assume that a graph having properties (a), (b), (c′) and fewer than $m > 0$ arcs is unicursal, and consider such a graph G_2 having m arcs. By (b) and (c′), the graph G_2 contains a closed route that visits more than one node. Delete the arcs of this closed route, yielding a graph G_2' that satisfies (b) and (c′), and consider the connected pieces of G_2'. Each piece satisfies (a), (b), and (c′), has fewer than m arcs, hence is unicursal by the inductive assumption. Since G_2 is connected, each of these unicursal pieces has at least one node in common with the closed route. It follows that G_2 is unicursal, as was to be shown.

It may also be observed that in application to any mixed graph satisfying (a) and (b), the simplest method of testing for unicursality is probably to set up the problem in flow form and attempt to construct a feasible circulation by the method of § 3. If there is no feasible circulation, then the method yields sets X and $\bar{X}$ for which the hypothesis (c) fails to be satisfied.

8. Dilworth's chain decomposition theorem for partially ordered sets

Let P be a finite partially ordered set with elements $1, 2, \ldots, n$ and order relation "$\succ$". A *chain* in P is a set of one or more elements $i_1, i_2, \ldots, i_k$ with

$$(8.1) \qquad\qquad i_1 \succ i_2 \succ \ldots \succ i_k.$$

(If we associate a directed graph with P by taking nodes $1, 2, \ldots, n$ and arcs (i, j) corresponding to $i \succ j$, this notion of a chain coincides with the notion of chain in the graph, except that now we allow a single node to be a chain.) A *decomposition* of P is a partition of P into chains. Thus P always has the trivial decomposition into n 1-element chains. A decomposition with the smallest number of chains is *minimal*.

Two distinct members i, j of P are *unrelated* if neither $i \succ j$ nor $j \succ i$. Notice that the maximal number of mutually unrelated elements of P is less than or equal to the number of chains in a minimal decomposition of P, since two members of a set of mutually unrelated elements cannot belong to the same chain. The finite case of Dilworth's chain decomposition theorem asserts that actually equality holds in the inequality just stated [3].

EXAMPLE. In the partially ordered set depicted in Fig. 8.1, all arcs are oriented downward and arcs corresponding to relations implied by transitivity have been omitted. The heavy lines indicate a decomposition into

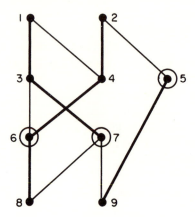

Figure 8.1

three chains and the three circled nodes are a set of mutually unrelated elements.

Dantzig and Hoffman have shown how to formulate the problem of finding a minimal decomposition of a partially ordered set as a linear programming problem, and have deduced Dilworth's theorem from duality theory [2]. Here we shall establish the connection between this theorem and the König-Egerváry theorem. It will follow that the problem of constructing a minimal decomposition can be solved by the algorithm of § 5 [7].

Given the partially ordered set $P = \{1, 2, \ldots, n\}$, let $G = [S, T; \mathscr{A}]$ be the bipartite graph consisting of $2n$ nodes $S = \{x_1, \ldots, x_n\}$, $T = \{y_1, \ldots, y_n\}$, and arcs defined from P by the rule: $(x_i, y_j) \in \mathscr{A}$ if and only if $i \succ j$. Using the language introduced in the array version of Theorem 4.1, we shall refer to independent sets of arcs and covering sets of nodes in G. A covering set of nodes is *proper* if no proper subset is itself a cover.

LEMMA 8.1. *Corresponding to any independent set $\mathscr{I} \subseteq \mathscr{A}$ there is a decomposition Δ of P with $|\mathscr{I}| + |\Delta| = n$.*

PROOF. Let

$$(8.2) \qquad \mathscr{I} = \{(x_{i_1}, y_{i_2}), (x_{i_3}, y_{i_4}), \ldots, (x_{i_{2k-1}}, y_{i_{2k}})\}.$$

Thus

$$(8.3) \qquad i_1 \succ i_2, i_3 \succ i_4, \ldots, i_{2k-1} \succ i_{2k}$$

in P, and we may group the distinct elements of the set $\{i_1, \ldots, i_{2k}\}$ into chains, each containing two or more elements. These chains are disjoint, since $\mathscr{I}$ is an independent set of arcs in G. By adding to these, as one

62

element chains, all elements of P that do not already appear, a decomposition Δ of P is obtained. If the number of elements of P that are in the jth chain of Δ is l_j, it follows that

$$(8.4) \qquad n = \sum_{j=1}^{|\Delta|} l_j = \sum_{j=1}^{|\Delta|} (l_j - 1) + |\Delta| = |\mathscr{I}| + |\Delta|,$$

since $l_j - 1$ counts the number of arcs of $\mathscr{I}$ that are used in forming the jth chain of Δ.

Notice that the proof of Lemma 8.1 does not make full use of the assumption that P is partially ordered. Indeed, Lemma 8.1. is valid for directed graphs that contain no directed cycles.

LEMMA 8.2. *Corresponding to any proper cover* $X \subseteq S \cup T$, *there is a set* $U \subseteq P$ *of mutually unrelated elements with* $|X| + |U| = n$.

PROOF. Let

$$(8.5) \qquad X = \{x_{i_1}, \ldots, x_{i_k}, y_{j_1}, \ldots, y_{j_m}\}$$

be a proper cover. The elements of the set of indices in (8.5) are distinct, for suppose $i_1 = j_1$, say. Since X is a proper cover, there is an $x_r \notin X$ with $(x_r, y_{j_1}) \in \mathscr{A}$; similarly there is a $y_s \notin X$ with $(x_{i_1}, y_s) \in \mathscr{A}$. Then, by transitivity and the assumption that $i_1 = j_1$, it follows that $(x_r, y_s) \in \mathscr{A}$. This contradicts the assumption that X is a cover, and thus implies that the elements of the set $\{i_1, \ldots, i_k, j_1, \ldots, j_m\}$ are all distinct. Now let U be the complement in P of this set. Since X is a cover, the elements of U are mutually unrelated, and $n = |X| + |U|$.

Dilworth's theorem now follows from the lemmas and Theorem 4.1. For let $\hat{\mathscr{I}}$ be a maximal independent set, $\hat{X}$ a minimal cover, and let $\hat{\Delta}$, $\hat{U}$ be their respective correspondents in P. By Theorem 4.1, $|\hat{\mathscr{I}}| = |\hat{X}|$; hence, by the lemmas, $|\hat{\Delta}| = |\hat{U}|$. But, as we have observed, $|U| \leqslant |\Delta|$ for all U and Δ.

It is true, conversely, that Dilworth's theorem implies Theorem 4.1. This can be seen by making the given bipartite graph $G = [S, T; \mathscr{A}]$ into a partially ordered set by defining, for $x \in S$, $y \in T$, the relation $x \succ y$ corresponding to $(x, y) \in \mathscr{A}$. The desired implication now follows from the following two easily checked statements:

(a) corresponding to any decomposition Δ of P, there is a set of independent arcs $\mathscr{I}$ of G with $|\Delta| + |\mathscr{I}| = |S \cup T|$; namely, let $\mathscr{I}$ be the two element chains of Δ;

(b) corresponding to any set $U \subseteq P$ of unrelated elements, there is a cover X of G with $|U| + |X| \leqslant |S \cup T|$, for the complement of U contains a cover.

From the proofs of the lemmas, it is clear that the algorithm described in § 5 can be used to construct a minimal chain decomposition and a maximal set of mutually unrelated elements.

EXAMPLE (continued). For the partially ordered set of Fig. 8.1, the equivalent array problem is schematized in Fig. 8.2. The assignment of 1's

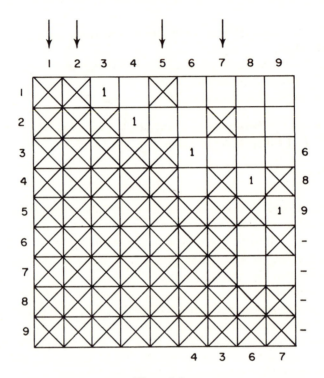

Figure 8.2

shown there, obtained using the starting procedure suggested in § 5, corresponds to the chain decomposition {1, 3, 6; 2, 4, 8; 5, 9; 7}. The labeling shown breaks through to column 7. After making the indicated switch, we obtain the new decomposition {1, 3, 7; 2, 4, 6, 8; 5, 9} containing one fewer chain. The next labeling produces labels on rows 5, 6, 7, 8, 9 and columns 8, 9 without breakthrough. Consequently the unlabeled rows 1, 2, 3, 4 and labeled columns 8, 9 are a minimal cover, and hence the "missing" elements 5, 6, 7 form a maximal mutually unrelated set.

9. Minimal number of individuals to meet a fixed schedule of tasks

As an application of the maximal flow computation for a minimal chain decomposition of a partially ordered set, consider the problem of finding the fewest number of individuals required to meet a fixed schedule of tasks.

Suppose there are n tasks T_i with stipulated starting times a_i and finishing times b_i $(a_i \leqslant b_i)$ and that reassignment times from T_i to T_j are given by numbers $r_{ij} \geqslant 0$ for $i \neq j$. How many individuals are needed to perform all tasks on schedule?

For example, the individuals might be machines of a given type, and r_{ij} might represent the set-up time necessary, having finished T_i, before T_j can be started. Or the individuals might be airplanes, say, the T_i scheduled flights, and r_{ij} the time required to return from the point of destination of flight T_i to the point of origin of flight T_j [1].

Making the reasonable assumption that

$$(9.1) \qquad\qquad r_{ik} \leqslant r_{ij} + r_{jk},$$

it is easy to see that the T_i can be partially ordered by defining

$$(9.2) \qquad T_i \succ T_j \text{ if and only if } b_i + r_{ij} \leqslant a_j,$$

and that, in terms of this partial ordering, a chain represents a possible assignment of tasks to one individual. Thus we are seeking a minimal chain decomposition of this partially ordered set.

It follows from Dilworth's theorem that the fewest number of individuals required is equal to the maximal number of tasks, no two of which can be performed by the same individual.

One special case in which a minimal chain decomposition can be found by a simple decision rule, without recourse to an iterative procedure like the labeling process, is that in which the elements of the partially ordered set can be numbered $1, \ldots, n$ in such a way that $i \leqslant j$ implies that the predecessors of i are included in those of j. In terms of the corresponding array, this is equivalent to saying that the rows and columns can be rearranged so that the set of inadmissible cells has "echelon" or "staircase" form (see Fig. 9.2, below). Assuming they have been so arranged, the following rule solves the problem.

Staircase rule. Select any admissible cell that borders the staircase of inadmissible cells and place a 1 in it. Delete the corresponding row and column and repeat the procedure. (Here "border" means "have a segment in common.")

Notice that the inadmissible cells of the reduced problem have staircase form, so the rule makes sense.

At termination of the process, a minimal cover can be found simply by selecting as many consecutive rows as possible that contain 1's (starting from the top), then switching to columns to cover the remaining admissible cells, if any.

In terms of the partially ordered set, the rule might be phrased as follows. Assuming that the set has been numbered as stipulated above, select an undominated element (e.g., element 1), then proceed to its first (in terms of the numbering) successor j, then to the first successor k of j, and so on until an element having no successor is reached. This traces out one chain of a minimal decomposition. Delete the elements of this chain and repeat the process.

EXAMPLE. In the partially ordered set shown in Fig. 9.1, or in array form in Fig. 9.2, the rule leads to the minimal decomposition {1, 2, 6, 7; 3, 4, 5}. A minimal covering in Fig. 9.2 consists of row 1 and columns 4, 5, 6, 7; this singles out 2 and 3 as a maximal set of unrelated elements.

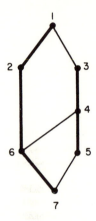

Figure 9.1

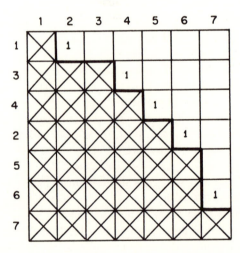

Figure 9.2

A proof that the rule works for problems having staircase form can be based on the fact that in any proper covering of the admissible cells for such a problem, an admissible cell bordering an inadmissible cell is covered by just one line. (For example, the proper coverings in Fig. 9.2 are: last six columns; first row, last four columns; first two rows, last three columns; first three rows, last two columns; first four rows, last column; first six rows.) We omit the details of a proof; the interested reader will have no difficulty constructing his own.

More difficult combinatorial problems along these lines emerge if the requirement that the schedule of tasks has been fixed in advance is dropped. For example, one might consider tasks $T_1, \ldots, T_n$ with known times $t_1, \ldots, t_n$ to perform each task, stipulated reassignment times r_{ij}, and pose either of the following questions:

(a) assuming that all tasks must be completed by time T, arrange a schedule that requires the fewest number of individuals;

(b) assuming a fixed number of individuals, arrange a schedule that completes all tasks at the earliest time.

The problem statement might be further complicated by the assumption of technological ordering restrictions—e.g., washing a dish precedes drying same—or, by the assumption of different types of individuals—e.g., dish washers and dish dryers. As far as we know, no computationally good ways of solving problems of this genre are known, although some special results have been deduced. For instance, Johnson has given an elegant solution to a problem of type (b) above in which there are two individuals, w (the washer) and d (the dryer), $2n$ tasks $W_1, \ldots, W_n, D_1, \ldots, D_n$, with the restrictions that W_i must precede D_i and that w specializes in W's, d in D's; the reassignment times from W_i to W_j or from D_i to D_j are assumed zero [20].

10. Set representatives

Let
$$\mathscr{S} = \{S_1, \ldots, S_n\}$$
be a family of subsets of a given set
$$E = \{e_1, \ldots, e_m\}.$$
A list R of distinct elements of E,
$$R = \{e_{i_1}, e_{i_2}, \ldots, e_{i_n}\}$$
is a *system of distinct representatives* (customarily abbreviated SDR) for $\mathscr{S}$ if
$$e_{i_j} \in S_j, \qquad\qquad j = 1, \ldots, n,$$
and e_{i_j} is said to *represent* S_j.

EXAMPLE 1. Let $E = \{1, 2, 3, 4, 5\}$, and suppose $\mathscr{S}$ is composed of $S_1 = \{2, 4, 5\}$, $S_2 = \{1, 5\}$, $S_3 = \{3, 4\}$, $S_4 = \{3, 4\}$. Then $R = \{5, 1, 3, 4\}$ is an SDR for $\mathscr{S}$ in which 5 represents S_1, 1 represents S_2, and so on.

EXAMPLE 2. Let $S_1 = \{1, 2\}$, $S_2 = \{2\}$, $S_3 = \{2, 3, 4, 5\}$, $S_4 = \{1, 2\}$. Here there is no SDR, since S_1, S_2, S_4 contain between them only two elements.

EXAMPLE 3. Let the fundamental set E consist of all U.S. Senators, and let $S_1, \ldots, S_n$ be an enumeration of Senate Committees. Can one find n distinct Senators $e_{i_1}, \ldots, e_{i_n}$ such that Senator e_{i_j} is a member of Committee S_j?

EXAMPLE 4. Suppose there are m men and n women, and that woman j rates man i as matrimonially acceptable or unacceptable. When is it possible to contract n marriages so that each woman has a husband acceptable to her?

Necessary and sufficient conditions for the existence of an SDR are contained in the following well-known theorem of P. Hall [16].

THEOREM 10.1. *An SDR exists for* $\mathscr{S} = \{S_1, \ldots, S_n\}$ *if and only if every union of k sets of $\mathscr{S}$ contains at least k distinct elements, $k = 1, \ldots, n$.*

As in the flow feasibility theorems, the necessity of the Hall condition is of course obvious.

In this section we shall discuss some Hall-type theorems that can be deduced from the flow feasibility theorems presented earlier in the chapter. While each of these theorems can be regarded as a generalization of Hall's theorem, it is perhaps misleading to emphasize this point, since it is equally true that each can be deduced from Hall's theorem. Indeed, one can show that the max-flow min-cut theorem is a consequence of Hall's theorem; the proof is lengthy, but see [17], for example, where such a proof of the integral form of the circulation theorem is given.

Before proceeding to other set representative problems, we first give a flow proof of Hall's theorem. Specifically, we shall show that the sufficiency of Hall's condition is an immediate consequence of Corollary 1.2, the second version of the supply-demand theorem, and, of course, the integrity theorem. To see this, define the bipartite network $G = [S, T; \mathscr{A}]$ with

$$S = \{x_1, \ldots, x_m\}, \qquad T = \{y_1, \ldots, y_n\}, \qquad \mathscr{A} = \{(x_i, y_j) | e_i \in S_j\}.$$

Associate a demand $b(y_j) = 1$ with each node of T, a supply $a(x_i) = 1$ with each node of S. The capacities of all arcs may be taken infinite. (See Fig. 10.1.) It is then clear that an integral feasible flow from S to T picks out an SDR for $S_1, \ldots, S_n$, and conversely. (Since we are requiring only that $f(S, y_j) \geq 1$, the same set may be represented more than once; this poses no difficulties.)

If Hall's condition is satisfied, then for any subset $T' \subseteq T$, there are at

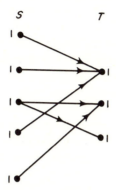

Figure 10.1

least $b(T') = |T'|$ nodes of S that are joined to nodes of T', and hence the flow $f_{T'}$ of Corollary 1.2 exists. Consequently the flow problem is feasible, and $\mathscr{S}$ admits an SDR.

As corollaries of Hall's theorem, we mention the following sufficient conditions for the existence of an SDR. Corollary 10.2 has been stated in [24].

COROLLARY 10.2. *Suppose* $S_j(j = 1, \ldots, n)$ *contains* s_j *elements, that* $e_i(i = 1, \ldots, m)$ *occurs in* r_i *of the sets* S_j, *and let* $S = \sum_{j=1}^{n} s_j = \sum_{i=1}^{m} r_i$, $M = \max(r_1, \ldots, r_m, s_1, \ldots, s_n)$. *If* $(n-1)M < S$, *then the family* $\mathscr{S} = \{S_1, \ldots, S_n\}$ *has an* SDR.

PROOF. Suppose Hall's condition is violated, so that there are k sets, say $S_1, \ldots, S_k$, which collectively contain $l < k$ elements. Then some one of these l elements must occur in p of the sets S_j, where $pl \geqslant \sum_1^k s_j$. By assumption, we have

$$S = \sum_1^k s_j + \sum_{k+1}^n s_j > (n-1)M,$$

and thus

$$pl > (n-1)M - \sum_{k+1}^n s_j \geqslant (n-1)M - (n-k)M = (k-1)M.$$

Since $l \leqslant k-1$, we must have $p > M$, a contradiction.

COROLLARY 10.3. *Suppose there are* n *elements* e_i *and* n *sets* S_j, *that each* S_j *contains* $k > 0$ *elements and each* e_i *occurs in* k *sets. Then* $\mathscr{S} = \{S_1, \ldots, S_n\}$ *has an* SDR.

PROOF. Immediate from Corollary 10.2.

A re-interpretation and repeated application of Corollary 10.3 yields:

69

COROLLARY 10.4. *If A is an n by n matrix composed of the integers 0 and 1 with $k > 0$ 1's in each row and column, then A is a sum of k permutation matrices.*

Suppose next that we drop the requirement concerning distinctness of representatives, and insist instead that each element $e_i \in E$ must occur in the system of representatives R at least a_i times and at most a'_i times, where $0 \leqslant a_i \leqslant a'_i$. (Thus if $a_i = 0$, $a'_i = 1$, for all $i = 1, \dots, m$, this reduces to the SDR problem.) We term R a *system of restricted representatives* (SRR). In matrimonial terms, the problem is now a polygynous one in which man i requires a_i wives, but can handle at most a'_i wives.

The following theorem gives existence conditions for an SRR [6].

THEOREM 10.5. *An SRR in which e_i occurs at least a_i, and at most a'_i, times $(0 \leqslant a_i \leqslant a'_i)$, exists for $\mathscr{S} = \{S_1, \dots, S_n\}$ if and only if, for every subset X of the indices $\{1, \dots, n\}$,*

$$(10.1) \qquad |X| \leqslant \min \left(n - \sum_1^m a_i + \sum_{I(X)} a_i, \ \sum_{I(X)} a'_i \right).$$

Here $I(X) \subseteq \{1, \dots, m\}$ is the index set of $\bigcup_{j \in X} S_j$.

PROOF. The conditions (10.1) are perhaps most easily discovered by applying the symmetric supply-demand theorem to the network used in the proof of Hall's theorem. This time we insist that the total flow out of each node $x_i \in S$ $(i = 1, \dots, m)$ be at least $a(x_i) = a_i$ and at most $a'(x_i) = a'_i$, and the flow into $y_j \in T$ be precisely one, $b(y_j) = b'(y_j) = 1$. Again arc capacities are infinite. Thus integral feasible flows and SRR's correspond by the rule: $f(x_i, y_j) = 1$ if and only if e_i represents S_j.

Let $X \subseteq S \cup T$ and define

$$S \cap X = U, \qquad S \cap \bar{X} = \bar{U}, \qquad T \cap X = V, \qquad T \cap \bar{X} = \bar{V},$$

so that $\bar{U}$ is the complement of U in S, and $\bar{V}$ is the complement of V in T. Then conditions (2.5) of Theorem 2.1 become

$$(10.2) \qquad c(U, \bar{V}) \geqslant |\bar{V}| - a'(\bar{U}), \qquad \text{all } U \subseteq S, V \subseteq T,$$

which holds automatically unless $(U, \bar{V}) = \varnothing$.

If we extend the notations $A(x)$, $B(x)$, introduced earlier (I.1), to subsets X of nodes in the usual fashion, e.g.,

$$B(X) = \bigcup_{x \in X} B(x),$$

then, for the case at hand, the statement $(U, \bar{V}) = \varnothing$ is equivalent to either of the statements $B(\bar{V}) \subseteq \bar{U}$ or $A(U) \subseteq V$. Using the former of these, we see that the set of inequalities (10.2) is equivalent to the set

$$(10.3) \qquad |\bar{V}| \leqslant a'(B(\bar{V})), \qquad \text{all } \bar{V} \subseteq T.$$

Similarly, conditions (2.6) of Theorem 2.1 become

$$c(U, \bar{V}) \geqslant a(U) - |V| = a(U) - n + |\bar{V}|$$

for all $U \subseteq S$, $\bar{V} \subseteq T$, or

$$|\bar{V}| \leqslant n - a(U) = n - a(S) + a(\bar{U})$$

for all $\bar{V} \subseteq T$, all $\bar{U}$ such that $B(\bar{V}) \subseteq \bar{U} \subseteq S$. Consequently the set of inequalities

(10.4) $$\qquad\qquad |\bar{V}| \leqslant n - a(S) + a(B(\bar{V})), \qquad\qquad \text{all } \bar{V} \subseteq T,$$

together with (10.3), are necessary and sufficient for the existence of (integral) feasible flows.

Translating (10.3) and (10.4) to set theoretic statements yields (10.1). Thus the inequalities

$$|X| \leqslant \sum_{I(X)} a_i'$$

are necessary and sufficient for the existence of an SRR in which each e_i can occur at most a_i' times, while

$$|X| \leqslant n - \sum_1^m a_i + \sum_{I(X)} a_i$$

are conditions for the existence of an SRR in which each e_i must occur at least a_i times.

Notice that taking $a_i = 0$, $a_i' = 1$ in (10.1) gives Hall's condition for the existence of an SDR.

In the special case that $a_i = 1$ for $i = 1, \ldots, l$, say, $a_i = 0$ for $i = l + 1$, $\ldots, m$ and $a_i' = 1$ all $i = 1, \ldots, m$, the SRR problem is that of determining conditions for the existence of an SDR containing the prescribed set of *marginal elements* $e_1, \ldots, e_l$. Mann and Ryser [23] proved that such an SDR exists if, in addition to Hall's condition, every marginal element appears at least t times among the S_j, where $t > 0$ is the largest number of marginal elements in any S_j. (They applied this sufficient condition to prove an interesting theorem due to Ryser [29] that provides necessary and sufficient conditions for extending an r by s Latin rectangle to an n by n Latin square.) The following theorem of Hoffman and Kuhn replaces the Mann-Ryser condition with a necessary and sufficient one [19].

THEOREM 10.6. *An* SDR *for* $\mathscr{S} = \{S_1, \ldots, S_n\}$ *that contains a prescribed set M of marginal elements exists if and only if both Hall's condition and the following condition hold : for any $M' \subseteq M$, the number of sets S_j that meet M' is at least $|M'|$.*

It is not difficult to see that the Mann-Ryser conditions imply the Hoffman-Kuhn conditions.

The validity of Theorem 10.6 can be seen from Hall's theorem and the general principle enunciated in Corollary 2.2. Applied here, Corollary 2.2 implies that we need only check the fact that the Hoffman-Kuhn condition is the feasibility condition for the problem obtained by interpreting the lower bounds $a(x_i)$ ($=1$ or 0 according as $x_i \in M$ or not) as demands in the reversed network, and the upper bounds $b'(y_j) = 1$ as supplies. But the Hoffman-Kuhn condition is precisely the Hall condition for feasibility in this situation.

Another necessary and sufficient condition for the marginal element problem that involves selections only of subsets of $\mathscr{S}$ is given directly by Theorem 10.5.

COROLLARY 10.7.   *An SDR for $\mathscr{S} = \{S_1, \ldots, S_n\}$ that contains a prescribed set $M = \{e_{i_1}, \ldots, e_{i_l}\}$ of marginal elements exists if and only if, for every $X \subseteq \{1, \ldots, n\}$,*

$$(10.5) \qquad |X| \leqslant \min \left(n - l + |L \cap I(X)|, |I(X)| \right).$$

Here $L = \{i_1, \ldots, i_l\}$ is the index set of M and $I(X)$ is the index set of $\bigcup_{j \in X} S_j$.

One can extend the SRR problem by asking for conditions under which a common SRR exists for two different collections

$$\mathscr{S} = \{S_1, \ldots, S_n\},$$
$$\mathscr{T} = \{T_1, \ldots, T_n\},$$

of subsets of the fundamental set, and still have a flow feasibility problem [6].

In matrimonial terms, the common SRR problem can be considered to have the following far-fetched interpretation. There are n men (corresponding to $S_1, \ldots, S_n$), n women (corresponding to $T_1, \ldots, T_n$) and m marriage brokers (corresponding to $e_1, \ldots, e_m$). Each broker has certain men and women clients, and must arrange at least a_i and at most a'_i marriages; all men and women must be married monogamously. When is this possible?

Here the network (see Fig. 10.2) may be taken as follows:

Nodes	Arcs		Arc lower bound function l	Arc capacity function c
s, t	(s, x_j)	$j = 1, \ldots, n$	0	1
$S = \{x_1, \ldots, x_n\}$	(x_j, y_i)	$\Leftrightarrow e_i \in S_j$	0	∞
$R = \{y_1, \ldots, y_m\}$	(y_i, y'_i)	$i = 1, \ldots, m$	a_i	a'_i
$R' = \{y'_1, \ldots, y'_m\}$	(y'_i, z_j)	$\Leftrightarrow e_i \in T_j$	0	∞
$T = \{z_1, \ldots, z_n\}$	(z_j, t)	$j = 1, \ldots, n$	0	1
	(t, s)		n	∞

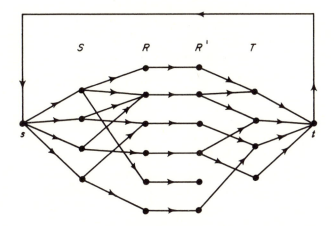

Figure 10.2

One sees easily that a common SRR exists for $\mathscr{S}, \mathscr{T}$ if and only if this network admits a feasible circulation; hence, by Theorem 3.1, if and only if the inequalities

$$c(X, \bar{X}) \geqslant l(\bar{X}, X)$$

hold for all subsets X of the nodes. To see what these inequalities reduce to for this particular network, let

$$S \cap X = U, \qquad S \cap \bar{X} = \bar{U},$$
$$R \cap X = V, \qquad R \cap \bar{X} = \bar{V},$$
$$R' \cap X = V', \qquad R' \cap \bar{X} = \bar{V}',$$
$$T \cap X = W, \qquad T \cap \bar{X} = \bar{W},$$

and begin by considering cases.

Case 1. $s \in X, t \in X$. Then

$$c(X, \bar{X}) = c(s, \bar{U}) + c(U, \bar{V}) + c(V, \bar{V}') + c(V', \bar{W}),$$
$$l(\bar{X}, X) = l(\bar{V}, V').$$

Case 2. $s \in X, t \in \bar{X}$. Then

$$c(X, \bar{X}) = c(s, \bar{U}) + c(U, \bar{V}) + c(V, \bar{V}') + c(V', \bar{W}) + c(W, t),$$
$$l(\bar{X}, X) = l(\bar{V}, V') + n.$$

Case 3. $s \in \bar{X}, t \in X$. Then $c(X, \bar{X})$ includes the term $c(t, s) = \infty$; consequently this case may be ignored.

Case 4. $s \in \bar{X}, t \in \bar{X}$. Then

$$c(X, \bar{X}) = c(U, \bar{V}) + c(V, \bar{V}') + c(V', \bar{W}) + c(W, t),$$
$$l(\bar{X}, X) = l(\bar{V}, V').$$

73

Next observe that the inequalities $c(X, \overline{X}) \geqslant l(\overline{X}, X)$ of *Case* 1 are dominated by those of *Case* 2, since $c(W, t) = |W| \leqslant n$. Similarly *Case* 4 inequalities are dominated by *Case* 2, since $c(s, \overline{U}) = |\overline{U}| \leqslant n$. Thus a feasible circulation exists if and only if

$$|\overline{U}| + c(U, \overline{V}) + c(V, \overline{V}') + c(V', \overline{W}) + |W| \geqslant l(\overline{V}, V') + n,$$

or

(10.6) $\quad |\overline{W}| + |U| \leqslant n + c(U, \overline{V}) + c(V, \overline{V}') + c(V', \overline{W}) - l(\overline{V}, V'),$

for all $U \subseteq S,\ V \subseteq R,\ V' \subseteq R',\ W \subseteq T$.

Again (10.6) is automatic unless the sets of arcs $(U, \overline{V})$ and $(V', \overline{W})$ are empty, that is, unless $A(U) \subseteq V$ and $B(\overline{W}) \subseteq \overline{V}'$. Then the right side of (10.6) is, if anything, decreased by taking $V = A(U)$ and $\overline{V}' = B(\overline{W})$. Thus, replacing $\overline{W}$ by W, (10.6) is equivalent to

(10.7) $\qquad |W| + |U| \leqslant n + c(A(U), B(W)) - l(\overline{A(U)}, \overline{B(W)})),$

for all $U \subseteq S,\ W \subseteq T$.

This proves

Theorem 10.8. *A common SRR in which* $e_i(i = 1, \ldots, m)$ *occurs at least* a_i *times and at most* a_i' *times* $(0 \leqslant a_i \leqslant a_i')$ *exists for* $\mathscr{S} = \{S_1, \ldots, S_n\}$ *and* $\mathscr{T} = \{T_1, \ldots, T_n\}$ *if and only if, for every* $X, Y \subseteq \{1, \ldots, n\}$,

(10.8) $\quad |X| + |Y| \leqslant n - \sum_{1}^{m} a_i + \sum_{I(X) \cup I(Y)} a_i + \sum_{I(X) \cap I(Y)} a_i'.$

Here $I(X) \subseteq \{1, \ldots, m\}$ *is the index set of* $\bigcup_{j \in X} S_j$ *and* $I(Y) \subseteq \{1, \ldots, m\}$ *that of* $\bigcup_{j \in Y} T_j$.

By taking $a_i = 0,\ a_i' = 1$ in Theorem 10.8, conditions for the existence of a common SDR are obtained.

Corollary 10.9. *A common SDR exists for* $\mathscr{S} = \{S_1, \ldots, S_n\}$ *and* $\mathscr{T} = \{T_1, \ldots, T_n\}$ *if and only if, for every* $X, Y \subseteq \{1, \ldots, n\}$,

(10.9) $\qquad\qquad |X| + |Y| \leqslant n + |I(X) \cap I(Y)|.$

Here $I(X)$ *is the index set of* $\bigcup_{j \in X} S_j$ *and* $I(Y)$ *that of* $\bigcup_{j \in Y} T_j$.

We conclude this section with the statement of one other set representative problem that can be solved as a flow feasibility problem: to find conditions for the existence of an SDR whose intersection with each member of a given partition of the fundamental set has cardinality between assigned bounds. This problem was posed and solved by Hoffman and Kuhn, who showed that it can be formulated as a linear program, and existence conditions established from duality theory [18].

THEOREM 10.10.   *Let a_k and a'_k, $k = 1, \ldots, p$, satisfying $0 \leqslant a_k \leqslant a'_k$, be integers associated with a given partition $P_1, \ldots, P_p$ of a given set $E = \{e_1, \ldots, e_m\}$. The subsets $S_1, \ldots, S_n$ of E have a system of distinct representatives R such that $a_k \leqslant |R \cap P_k| \leqslant a'_k$ if and only if*

$$(10.10) \quad \left| \left(\bigcup_{j \in U} P_k \right) \cap \left(\bigcup_{j \in V} S_j \right) \right| \geqslant \max \left(|V| - \sum_{\overline{U}} a'_k, \; |V| - n + \sum_{U} a_k \right)$$

holds for all $U \subseteq \{1, \ldots, p\}$ and $V \subseteq \{1, \ldots, n\}$.

A representing network for this problem may be taken as follows (see Fig. 10.3):

Nodes	Arcs	Arc capacities
$S = \{x_1, \ldots, x_p\}$	$(x_k, y_i) \Leftrightarrow e_i \in P_k$	1
$R = \{y_1, \ldots, y_m\}$	$(y_i, z_j) \Leftrightarrow e_i \in S_j$	∞
$T = \{z_1, \ldots, z_n\}$		

Associate with each $x_k \in S$ the bounds a_k, a'_k on flow out of x_k, with each $z_j \in T$ the bounds $b_j = b'_j = 1$ on flow into z_j. Theorem 10.10 then follows from the symmetric supply-demand theorem.

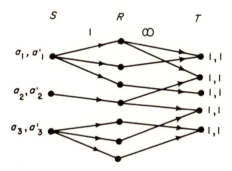

Figure 10.3

Thus, taking the first term on the right of (10.10), one has conditions for the existence of a system of distinct representatives R such that $|R \cap P_k| \leqslant a'_k$; the other term provides existence conditions for an R satisfying $|R \cap P_k| \geqslant a_k$.

11. The subgraph problem for directed graphs

Another graph theoretic combinatorial problem that can be dealt with by flow methods is one known as the subgraph problem for directed graphs. A solution to this problem has been given by Ore [25]. Various special cases have also received attention in the mathematical literature.

Here we shall use the symmetric supply-demand theorem to discuss a slight extension of this problem.

Let $G = [N; \mathscr{A}]$ be a finite directed graph, and let $e(x)$ and $i(x)$ be, respectively, the number of arcs entering and the number of arcs issuing from node x. Then the (*local*) *degree* of G at x is the pair $e(x)$, $i(x)$. The subgraph problem for G is that of determining conditions under which G has a subgraph H having prescribed local degrees. Here a *subgraph H of G* is a graph $H = [N; \mathscr{A}']$ with $\mathscr{A}' \subseteq \mathscr{A}$.

Consider the following generalization of this problem. Associate with each $x \in N$ four integers

$$a(x), \quad a'(x), \quad b(x), \quad b'(x)$$

satisfying

(11.1) $$0 \leqslant a(x) \leqslant a'(x),$$

(11.2) $$0 \leqslant b(x) \leqslant b'(x),$$

and find conditions under which G has a subgraph H whose local degrees $e_H(x)$, $i_H(x)$ satisfy

(11.3) $$a(x) \leqslant i_H(x) \leqslant a'(x),$$

(11.4) $$b(x) \leqslant e_H(x) \leqslant b'(x).$$

To determine such conditions, we convert the problem to a flow problem and apply Theorem 2.1. First construct from G a bipartite graph $G' = [S, T; \mathscr{A}']$ having twice as many nodes as G but the same number of arcs: to each $x \in N$ correspond two nodes, $x' \in S$, $x'' \in T$; if $(x, y) \in \mathscr{A}$, then $(x', y'') \in \mathscr{A}'$, and these are all the arcs of G' (see Fig. 11.1). Assign unit capacity to each arc of G', and insist that the flow out of $x' \in S$ lie between $a(x)$ and $a'(x)$, the flow into $x'' \in T$ lie between $b(x)$ and $b'(x)$.

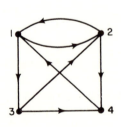 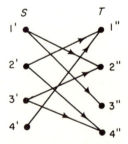

Figure 11.1

It is clear that an integral feasible flow f from S to T in G' yields a subgraph H of G satisfying (11.3), (11.4) by putting (x, y) in H if and only if $f(x', y'') = 1$, and conversely. Hence, letting U, V be arbitrary subsets

76

of S, T, respectively, and denoting their respective complements in S, T by $\bar{U}$, $\bar{V}$, it follows from Theorem 2.1 that H exists if and only if

$$(11.5) \qquad a'(\bar{U}) + |(U, \bar{V})| \geqslant b(\bar{V}),$$

$$(11.6) \qquad b'(V) + |(U, \bar{V})| \geqslant a(U),$$

hold for all $U \subseteq S$, $V \subseteq T$.

Before proceeding further, let us consider inequalities (11.5) for $a(x) = a'(x)$, $b(x) = b'(x)$, that is, in the case for which the local degrees of H are specified exactly. Then a necessary condition for H to exist is that $a(N) = b(N)$, or, in G',

$$(11.7) \qquad a(S) = b'(T).$$

On the other hand, (11.7) and (11.6) now imply (11.5), since

$$a(\bar{U}) + |(U, \bar{V})| \geqslant a(\bar{U}) + a(U) - b'(V) = a(S) - b'(V)$$
$$\geqslant b'(T) - b'(V) = b'(\bar{V}),$$

which is (11.5) with $a' = a$, $b' = b$.

Thus (11.7) and (11.6) are necessary and sufficient for the existence of a subgraph H having local degrees $e_H(x) = b'(x)$, $i_H(x) = a(x)$.

Each of the conditions (11.5), (11.6) is stated in terms of selections of pairs of sets. Each can, however, be simplified to a condition involving the choice of but one set. Consider (11.6), for example. For given $U \subseteq S$, let

$$V = \{y'' \in T \,|\, b'(y'') < |(U, y'')|\}.$$

For this pair U, V, the left-hand side of (11.6) may be written as

$$\sum_{y'' \in A(U)} \min\,[b'(y''), |(U, y'')|].$$

On the other hand, for fixed $U \subseteq S$, this sum minimizes $b'(V) + |(U, \bar{V})|$ over all $V \subseteq T$. Thus inequalities (11.6) are equivalent to the inequalities

$$(11.8) \qquad \sum_{y'' \in A(U)} \min\,[b'(y''), |(U, y'')|] \geqslant a(U), \qquad \text{all } U \subseteq S.$$

Similarly, (11.5) reduces to

$$(11.9) \qquad \sum_{y' \in B(\bar{V})} \min\,[a'(y'), |(y', \bar{V})|] \geqslant b(\bar{V}), \qquad \text{all } \bar{V} \subseteq T.$$

Thus, translating (11.8) and (11.9) to conditions stated in terms of the given graph G, we have the following theorem.

THEOREM 11.1.   Let $G = [N; \mathscr{A}]$ be a directed graph and suppose that, corresponding to each $x \in N$, there are integers $a(x)$, $a'(x)$, $b(x)$, $b'(x)$ with

$$(11.10) \qquad 0 \leqslant a(x) \leqslant a'(x),$$

$$(11.11) \qquad 0 \leqslant b(x) \leqslant b'(x).$$

Then G has a subgraph H whose local degrees $e_H(x)$, $i_H(x)$ satisfy

(11.12) $$a(x) \leqslant i_H(x) \leqslant a'(x),$$

(11.13) $$b(x) \leqslant e_H(x) \leqslant b'(x),$$

if and only if, for all $X \subseteq N$, we have

(11.14) $$a(X) \leqslant \sum_{y \in A(X)} \min [b'(y), |(X, y)|],$$

(11.15) $$b(X) \leqslant \sum_{y \in B(X)} \min [a'(y), |(y, X)|].$$

In view of the remark that (11.6) and (11.7) are necessary and sufficient for the existence of a subgraph H having prescribed local degrees, we may also state a theorem of Ore [25].

COROLLARY 11.2.   *The directed graph $G = [N; \mathscr{A}]$ has a subgraph H with local degrees*

(11.16) $$i_H(x) = a(x) \geqslant 0,$$

(11.17) $$e_H(x) = b(x) \geqslant 0,$$

if and only if

(11.18) $$a(N) = b(N),$$

and, for all $X \subseteq N$,

(11.19) $$a(X) \leqslant \sum_{y \in A(X)} \min [b(y), |(X, y)|].$$

Notice also, as a consequence of Corollary 2.2, that if G has subgraphs H_1, H_2 such that

$$b(x) \leqslant e_{H_1}(x), \qquad i_{H_1}(x) \leqslant a'(x),$$

$$e_{H_2}(x) \leqslant b'(x), \qquad a(x) \leqslant i_{H_2}(x),$$

then G has a subgraph H such that

$$a(x) \leqslant i_H(x) \leqslant a'(x),$$

$$b(x) \leqslant e_H(x) \leqslant b'(x).$$

For undirected graphs G, the (local) degree of G at x is the number of arcs incident with x, and the subgraph problem is that of determining conditions under which G has a subgraph H with specified local degrees.

This problem has been solved by Tutte [34], and also by Ore [26]. In contrast with the directed case, we know of no formulation of the undirected problem as a flow feasibility problem.†

12. Matrices composed of 0's and 1's

An m by n matrix whose entries are the integers 0 and 1 can be thought of as distributing n elements into m sets: the 1's in row i designate the elements that occur in the ith set, and the 1's in column j designate the sets that contain the jth element. In other words, the matrix may be thought of as an incidence matrix of sets *versus* elements. Such matrices may thus be regarded as fundamental in combinatorial investigations.

Ryser has focused attention on the class of all m by n (0, 1)-matrices having prescribed row and column sums, and has obtained a number of results that give insight into combinatorial properties of the class [30, 31, 32]. Some of these results are accessible through the use of network flows; others appear not to be.

The first question that naturally arises for such matrices is: when do they exist? That is, given non-negative integers $a_1, \ldots, a_m$, and $b_1, \ldots, b_n$, when does there exist an m by n (0, 1)-matrix having a_i 1's in the ith row and b_j 1's in the jth column? Both Ryser [30] and Gale [11] independently answered this question. Not only do existence conditions here turn out to be much fewer in number than one might expect, but also a simple rule can be stated for constructing such a matrix.

Suppose we pose the existence problem in inequalities form, asking for a (0, 1)-matrix $A = (a_{ij})$ such that

$$(12.1) \qquad \sum_{i=1}^{m} a_{ij} \geqslant b_j,$$

$$(12.2) \qquad \sum_{j=1}^{n} a_{ij} \leqslant a_i.$$

† One could, of course, attempt to formulate the problem as a linear program of more general type. A way that suggests itself is to associate a variable bounded between 0 and 1 with each arc of the graph, impose the restriction for each node x that the sum of all variables corresponding to arcs incident with x should not exceed the specified degree of H at x, and ask for a feasible solution that maximizes the sum of all variables. The difficulty here, however, is that fractional solutions to the program can be obtained if the given graph G has odd cycles; that is, the important integrity property of maximizing solutions has been lost.

Of course the fact that a first naive statement of the problem leads to fractional solutions does not in itself imply that other linear programming formulations might not be useful. For instance, it can be shown that the convex polyhedron of flows in arc-chain form (I.2) has fractional vertices, but this is not true when the problem is put in node-arc form. Similarly, an attempt to pose the minimal chain decomposition problem for partially ordered sets in terms of the node-chain incidence matrix of the corresponding directed graph runs into the difficulty of fractional extreme solutions. But useful information about the combinatorial problem can be obtained from a different formulation, as we have seen.

In case $\sum a_i = \sum b_j$, the problem is that of filling an m by n array with 0's and 1's so that the ith row sum is a_i, and the jth column sum is b_j.

For a concrete example, consider n families to be seated at m tables, where the jth family has b_j members and the ith table a_i seats, in such a way that no two members of the same family are seated at the same table.

The existence problem (12.1), (12.2) for $(0, 1)$-matrices can be treated as a flow feasibility problem by setting up the bipartite network consisting of nodes

$$S = \{x_1, \ldots, x_m\}, \qquad T = \{y_1, \ldots, y_n\},$$

and arcs

$$\mathscr{A} = \{(x_i, y_j)\}, \qquad i = 1, \ldots, m; j = 1, \ldots, n.$$

Associate a demand $b(y_j) = b_j$ with the jth node of T, a supply $a(x_i) = a_i$ with the ith node of S, and impose unit capacity on all arcs. Then feasible integral flows $f(x_i, y_j)$ and $(0, 1)$-matrices (a_{ij}) satisfying (12.1) and (12.2) correspond via $a_{ij} = f(x_i, y_j)$.

If we use the second version of the supply-demand theorem (Corollary 1.2), we need to determine, corresponding to each $T' \subseteq T$, a flow $f_{T'}$ that maximizes $f(S, T')$, subject to the supply limitations at the sources S. Now $f(S, T')$ is maximized simply by sending as much as possible from each $x_i \in S$ to T'. Thus, if $|T'| = k$,

$$f_{T'}(x_i, T') = \min(a_i, k)$$

and

$$(12.3) \qquad f_{T'}(S, T') = \sum_{i=1}^{m} \min(a_i, k).$$

Suppose we picture the integers a_i as represented by rows of dots, for example:

$$
\begin{array}{cccccc}
a_1 & \cdot & \cdot & \cdot & \cdot & \cdot \\
a_2 & \cdot & \cdot \\
a_3 & \cdot & \cdot & \cdot \\
a_4 & \cdot \\
\end{array}
$$

Let a_j^* denote the number of dots in the jth column of the pictorial representation. Thus the sequences (a_i) and (a_j^*) are conjugate partitions of the integer $\sum a_i$, that is, a_j^* is the number of a_i that are greater than or equal to j. Then it is apparent that

$$(12.4) \qquad \sum_{i=1}^{m} \min(a_i, k) = \sum_{j=1}^{k} a_j^*.$$

Thus the problem is feasible if and only if

$$(12.5) \qquad b(T') \leqslant \sum_{j=1}^{|T'|} a_j^*$$

holds for all $T' \subseteq T$. Since the right side of (12.5) depends only on the cardinality of T', we may replace all the inequalities obtained by letting T' range over subsets of k elements with just one: that obtained by selecting T' to maximize $b(T')$. Thus (12.5), for all $T' \subseteq T$, is equivalent to the n inequalities

$$(12.6) \qquad \sum_{j=1}^{k} b_j \leqslant \sum_{j=1}^{k} a_j^*, \qquad\qquad k = 1, \ldots, n,$$

where we have chosen the notation so that $b_1 \geqslant b_2 \geqslant \ldots \geqslant b_n$.

This establishes the following theorem [11, 30].

THEOREM 12.1 *Let* a_i ($i = 1, \ldots, m$) *and* b_j ($j = 1, \ldots, n$) *be two sets of non-negative integers, where* $b_1 \geqslant b_2 \geqslant \ldots \geqslant b_n$. *Then there is an* m *by* n *(0, 1)-matrix* $A = (a_{ij})$ *satisfying*

$$(12.7) \qquad \sum_{i} a_{ij} \geqslant b_j, \qquad \sum_{j} a_{ij} \leqslant a_i,$$

if and only if

$$(12.8) \qquad \sum_{j=1}^{k} b_j \leqslant \sum_{j=1}^{k} a_j^*, \qquad\qquad k = 1, \ldots, n,$$

where $a_j^* = |\{i \,|\, a_i \geqslant j\}|$.

Easy consequences of Theorem 12.1 are

COROLLARY 12.2. *If* $b_j = b$ *for all* j, *there is a* (0, 1)-*matrix satisfying* (12.7) *if and only if* $nb \leqslant \sum_{j=1}^{n} a_j^*$.

COROLLARY 12.3. *If* $a_i = a$ *for all* i, *there is a* (0, 1)-*matrix satisfying* (12.7) *if and only if* $\sum_{j=1}^{n} b_j \leqslant ma$, $b_j \leqslant m$.

PROOF OF COROLLARY 12.2. We need to show that $kb \leqslant \sum_1^k a_j^*$ for $k = 1, \ldots, n - 1$. Now $knb \leqslant k \sum_1^n a_j^*$ by assumption, so it suffices to establish $k \sum_1^n a_j^* \leqslant n \sum_1^k a_j^*$, which is equivalent to $k \sum_{k+1}^n a_j^* \leqslant (n - k) \sum_1^k a_j^*$. This inequality follows from the fact that the sequence (a_j^*) is monotone decreasing, since

$$k \sum_{k+1}^{n} a_j^* \leqslant k(n - k)a_{k+1}^* \leqslant (n - k)ka_k^* \leqslant (n - k) \sum_{1}^{k} a_j^*.$$

PROOF OF COROLLARY 12.3. The inequalities

$$\sum_{1}^{k} b_j \leqslant \sum_{1}^{k} a_j^* = \sum_{1}^{m} \min(a, k) = m \min(a, k)$$

follow from $\sum_1^n b_j \leqslant ma$ and $b_j \leqslant m$. For if $k \leqslant a$, then $\sum_1^k b_j \leqslant mk = m \min(a, k)$; if, on the other hand, $k > a$, then $\sum_1^k b_j \leqslant \sum_1^n b_j \leqslant ma = m \min(a, k)$.

Although we have used the second version of the supply-demand theorem in deriving the existence conditions (12.8), we might just as well have used the first version, or applied the max-flow min-cut theorem directly. The resulting existence conditions are worth stating explicitly, since they involve the "structure matrix" that has been introduced by Ryser in the study of $(0, 1)$-matrices having specified row and column sums [32]. Ryser has defined the structure matrix (t_{kl}) for the class $\mathfrak{A} = \mathfrak{A}(a_1, \ldots, a_m; b_1, \ldots, b_n)$ of all m by n $(0, 1)$-matrices having monotone decreasing row and column sums a_i and b_j to be the $m + 1$ by $n + 1$ matrix

$$t_{kl} = kl + a_{k+1} + \ldots + a_m - (b_1 + \ldots + b_l),$$
$$(12.9) \qquad\qquad k = 0, 1, \ldots, m; l = 0, 1, \ldots, n.$$

If the class $\mathfrak{A}$ is non-empty, then it can be seen directly that the t_{kl} are non-negative integers. For we may select a matrix A in $\mathfrak{A}$ and partition it thus:

$$(12.10) \qquad\qquad \begin{bmatrix} A_1 & * \\ * & A_2 \end{bmatrix}.$$

Here A_1 is k by l. It follows that t_{kl} defined by (12.9) is equal to the number of 0's in A_1 plus the number of 1's in A_2. It is true, conversely, that the non-negativity of the structure matrix implies that the set $\mathfrak{A}$ is non-empty. To see this, one can use the first supply-demand theorem. Applied here, this theorem asserts that $\mathfrak{A}$ is non-empty if the inequalities

$$(12.11) \qquad\qquad \sum_{\bar{J}} b_j - \sum_{\bar{I}} a_i \leqslant |I| \, |\bar{J}|$$

hold for all selections of subsets $I \subseteq \{1, \ldots, m\}$, $J \subseteq \{1, \ldots, n\}$. But for I and $\bar{J}$ of fixed cardinalities k and l, respectively, the left side of (12.11) is maximized by selecting $\bar{J} = \{1, \ldots, l\}$, $\bar{I} = \{k + 1, \ldots, m\}$, in view of the assumptions $a_1 \geqslant a_2 \geqslant \ldots \geqslant a_m, b_1 \geqslant b_2 \geqslant \ldots \geqslant b_n$. Then (12.11) is the statement that the entries of the structure matrix for the class $\mathfrak{A}$ are non-negative.

There is a simple, direct n-stage rule for constructing a $(0, 1)$-matrix satisfying the row and column sum constraints (12.7) in case the problem is feasible [10, 11]. Suppose that conditions (12.8) are satisfied and that we assign the ones in column p in some arbitrary fashion, say to the subset of rows $I = \{i_1, \ldots, i_{b_p}\}$. Let $\bar{a}_i$, $i = 1, \ldots, m$, and $\bar{b}_j$, $j = 1, \ldots, n - 1$, denote the upper and lower bounds on row and column sums in the reduced problem, so that

$$\bar{a}_i = \begin{cases} a_i - 1, & \text{if } i \in I, \\ a_i, & \text{otherwise}, \end{cases}$$

$$\bar{b}_j = \begin{cases} b_j, & j = 1, \ldots, p - 1, \\ b_{j+1}, & j = p, \ldots, n - 1. \end{cases}$$

By Theorem 12.1, the reduced problem is feasible if and only if

$$(12.12) \qquad \sum_{j=1}^{k} \bar{b}_j \leqslant \sum_{j=1}^{k} \bar{a}_j^*, \qquad k = 1, \ldots, n-1,$$

where $(\bar{a}_j^*)$ is the conjugate sequence to $(\bar{a}_j)$. Now the right side of (12.12) can be rewritten as

$$\sum_{1}^{k} a_j^* - b_p + a_{k+1}^*(I),$$

where $a_{k+1}^*(I)$ is the number of a_i such that $i \in I$ and $a_i \geqslant k+1$. Consequently the feasibility conditions for the reduced problem are

$$(12.13) \qquad \sum_{j=1}^{k} b_j + b_p \leqslant \sum_{j=1}^{k} a_j^* + a_{k+1}^*(I), \quad k = 1, \ldots, p-1,$$

$$(12.14) \qquad \sum_{j=1}^{k+1} b_j \leqslant \sum_{j=1}^{k} a_j^* + a_{k+1}^*(I), \qquad k = p, \ldots, n-1.$$

If we specialize I to correspond to the b_p largest a_i, then

$$a_{k+1}^*(I) = \min(b_p, a_{k+1}^*),$$

and conditions (12.13), (12.14) always hold under the assumption of feasibility for the original problem. For if $k < p$ and $\min(b_p, a_{k+1}^*) = b_p$, (12.13) becomes

$$\sum_{j=1}^{k} b_j \leqslant \sum_{j=1}^{k} a_j^*;$$

if $k < p$ and $\min(b_p, a_{k+1}^*) = a_{k+1}^*$, (12.13) becomes

$$\sum_{j=1}^{k} b_j + b_p \leqslant \sum_{j=1}^{k+1} a_j^*,$$

which is valid since $b_1 \geqslant b_2 \geqslant \ldots \geqslant b_n$. If, on the other hand, $k \geqslant p$ and $\min(b_p, a_{k+1}^*) = b_p$, (12.14) reduces to

$$\sum_{\substack{j=1 \\ j \neq p}}^{k+1} b_j \leqslant \sum_{j=1}^{k} a_j^*,$$

again a valid inequality since the b_j are monotone decreasing; if $k \geqslant p$ and $\min(b_p, a_{k+1}^*) = a_{k+1}^*$, (12.14) becomes

$$\sum_{j=1}^{k+1} b_j \leqslant \sum_{j=1}^{k+1} a_j^*.$$

Thus the following rule either constructs a solution or shows that the problem is infeasible.

(0, 1)-*matrix rule. Select any column, assign its 1's to the rows having largest row sum bounds, and repeat the procedure in the reduced problem.*

In terms of the table seating problem, all n families can be seated in n stages by selecting, at the jth stage, any family not already seated, and distributing its members among those tables having the most vacant seats.

EXAMPLE.

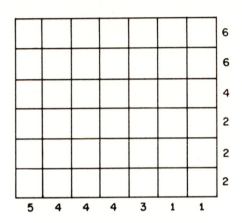

(conjugate sequence)

The feasibility conditions of Theorem 12.1 hold, since

$$5 \leqslant 6,$$
$$9 \leqslant 12,$$
$$13 \leqslant 15,$$
$$17 \leqslant 18,$$
$$20 \leqslant 20,$$
$$21 \leqslant 22,$$
$$22 \leqslant 22.$$

84

Using the rule, the following solution is found:

1	1	1	1	1	1		6
1	1	1	1	1		1	6
1	1	1	1				4
1		1					2
1			1				2
	1			1			2
5	4	4	4	3	1	1	

We point out, in view of Corollary 2.3, that if there are two $(0, 1)$-matrices such that one satisfies upper bounds on row sums, lower bounds on column sums, while the other satisfies lower bounds on row sums, upper bounds on column sums, then there is a $(0, 1)$-matrix satisfying the designated lower and upper bounds on both row and column sums, provided, of course, that the lower bounds on rows (columns) do not exceed the upper bounds on rows (columns).

Assuming $m = n$, the existence problem for $(0, 1)$-matrices can also be interpreted as one concerning the existence of directed graphs on n nodes having specified local degrees, where we now permit circular arcs, that is, arcs that lead from a node to itself. Here an entry 1 in the ij position means there is an arc from i to j. If we do not allow circular arcs, the problem becomes one of filling an n by n matrix with 0's and 1's, subject to stated upper bounds a_i on row sums, lower bounds b_j on column sums, and the added restriction that 1's cannot be placed along the main diagonal. In other words, we require that the trace of the matrix be zero [9].

Following the same procedure used in the proof of Theorem 12.1, it is not difficult to see that feasibility conditions for this latter problem can be stated as

$$(12.15) \qquad \sum_I b_i \leqslant \sum_1^{|I|-1} a_i^* + a_{|I|}^*(\bar{I}), \qquad \text{all } I \subseteq \{1, \ldots, n\},$$

where $a_{|I|}^*(\bar{I})$ is the number of a_i such that $a_i \geqslant |I|$ and $i \in \bar{I}$. These inequalities simplify considerably if we make the assumption that there is a common rearrangement of the a's and b's such that

$$(12.16) \qquad a_1 \geqslant a_2 \geqslant \ldots \geqslant a_n,$$

$$(12.17) \qquad b_1 \geqslant b_2 \geqslant \ldots \geqslant b_n.$$

Under these circumstances,

$$\sum_I b_i - a_{|I|}^*(I)$$

is maximized, for $|I| = k$, by selecting $I = \{1, \ldots, k\}$. Thus (12.15) may be replaced by the n inequalities

(12.18) $$\qquad \sum_1^k b_i \leqslant \sum_1^{k-1} a_i^* + a_k^*(\{k + 1, \ldots, n\}), \qquad k = 1, \ldots, n.$$

Particular cases under which the common re-numbering (12.16), (12.17) exists are those corresponding to Corollaries 12.2 and 12.3, that is, $b_i = b$, all i, or $a_i = a$, all i.

If, instead of using the conjugate sequence (a_i^*), we push the dots representing the integers a_i as far as possible to the left, but this time place no dots in the main diagonal, e.g.,

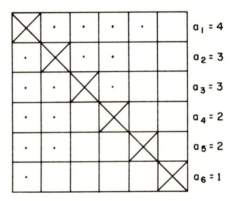

and define a_i^{**} to be the number of dots in the ith column, then

(12.19) $$\qquad \sum_1^{k-1} a_i^* + a_k^*(\{k + 1, \ldots, n\}) = \sum_{i=1}^k a_i^{**}.$$

Thus (12.18) simply asserts that the partial sums of the b-sequence are dominated by those of the a^{**}-sequence, and we have conditions analogous to those found by Ryser and Gale for the $(0, 1)$-matrix problem.

We shall refer to the sequence (a_i^{**}) as the *diagonally restricted conjugate sequence* in the following theorem, which summarizes this discussion.

THEOREM 12.4. *Let $a_1 \geqslant a_2 \geqslant \ldots \geqslant a_n$, $b_1 \geqslant b_2 \geqslant \ldots \geqslant b_n$ be non-negative integers. There is a $(0, 1)$-matrix (a_{ij}) satisfying*

(12.20) $$\qquad \sum_j a_{ij} \leqslant a_i, \qquad \sum_i a_{ij} \geqslant b_j, \qquad \sum_i a_{ii} = 0,$$

86

if and only if

(12.21)
$$\sum_{i=1}^{k} b_i \leqslant \sum_{i=1}^{k} a_i^{**}, \qquad k = 1, \ldots, n.$$

Here the sequence (a_i^{**}) *is the diagonally restricted conjugate of the sequence* (a_i).

The diagonally restricted conjugate sequence is not necessarily monotone, but it is enough so to make the analogue of Corollary 12.2 valid.

COROLLARY 12.5.   *If* $b_i = b$ *all* $i = 1, \ldots, n$, *then there is a* (0, 1)-*matrix satisfying* (12.20) *if and only if* $nb \leqslant \sum_{i=1}^{n} a_i^{**}$.

COROLLARY 12.6.   *If* $a_i = a$ *all* $i = 1, \ldots, n$, *then there is a* (0, 1)-*matrix satisfying* (12.20) *if and only if* $\sum_{i=1}^{n} b_i \leqslant na$, $b_i \leqslant n - 1$.

PROOF OF COROLLARY 12.5.   First note that

$$a_k^{**} = |I_k| + |J_k|,$$

where

$$I_k = \{i \,|\, i < k \quad \text{and} \quad a_i \geqslant k - 1\},$$
$$J_k = \{i \,|\, i > k \quad \text{and} \quad a_i \geqslant k\}.$$

We shall show that the monotonicity of the sequence (a_i) implies that either the sequence (a_i^{**}) is monotone, or else it has at most one point of increase, and that increase is one; that is, either

$$a_1^{**} \geqslant a_2^{**} \geqslant \ldots \geqslant a_n^{**},$$

or, for some $k = 1, \ldots, n - 1$,

$$a_1^{**} \geqslant \ldots \geqslant a_k^{**}, a_{k+1}^{**} = 1 + a_k^{**} \geqslant a_{k+2}^{**} \geqslant \ldots \geqslant a_n^{**}.$$

To see this, observe that

$$|I_k| \geqslant |I_{k+1}| - 1,$$
$$|J_k| \geqslant |J_{k+1}|,$$

and hence $a_k^{**} \geqslant a_{k+1}^{**} - 1$, equality holding if and only if equality holds in both the last displayed inequalities. Since $a_1 \geqslant a_2 \geqslant \ldots \geqslant a_n$, it follows that $a_k^{**} = a_{k+1}^{**} - 1$ if and only if $a_k \geqslant k$ and $a_{k+1} < k$. Thus, if there were two points of increase in the sequence (a_i^{**}), say k and l with $k < l$, then we should have

$$a_l \geqslant l > k > a_{k+1},$$

contradicting $a_{k+1} \geqslant a_l$. This completes the proof of the assertion made at the beginning of this paragraph.

To finish proving the corollary, we need to show that the inequalities

$$kb \leqslant \sum_{1}^{k} a_i^{**}, \qquad k = 1, \ldots, n,$$

follow from $nb \leqslant \sum_1^n a_i^{**}$ and the "almost monotone" property of the sequence (a_i^{**}). This can be established by induction on n, as follows. For $n = 1$, there is nothing to prove. Assume the proposition for $n - 1$, and consider the case for n. If $b < a_n^{**}$, then the almost monotone property, together with the fact that we are dealing with integers, implies that $b \leqslant a_i^{**}$ for all i. Consequently $kb \leqslant \sum_1^k a_i^{**}$. If, on the other hand, $b \geqslant a_n^{**}$, then we have $(n - 1)b \leqslant \sum_1^{n-1} a_i^{**}$, and the induction hypothesis applies.

PROOF OF COROLLARY 12.6. To show that $a_i = a$, $\sum_{i=1}^n b_i \leqslant na$, $b_i \leqslant n - 1$ imply $\sum_{i=1}^k b_i \leqslant \sum_1^k a_i^{**}$, where $b_1 \geqslant b_2 \geqslant \ldots \geqslant b_n$, consider two cases. If $k \leqslant a$, then $a_i^{**} = n - 1$ for $i \leqslant k$, and hence

$$\sum_1^k b_i \leqslant k(n - 1) = \sum_1^k a_i^{**}.$$

If, on the other hand, $k > a$, then

$$a_i^{**} = \begin{cases} n - 1 & \text{for } i \leqslant a, \\ a & \text{for } i = a + 1, \\ 0 & \text{for } i > a + 1, \end{cases}$$

and hence

$$\sum_1^k b_i \leqslant \sum_1^n b_i \leqslant na = \sum_1^k a_i^{**}.$$

We turn now from existence problems for $(0, 1)$-matrices having stated row and column sums to a brief discussion of other results concerning such matrices. Throughout this discussion we let $\mathfrak{A}$ denote the class of $(0, 1)$-matrices A having row sums a_i, column sums b_j, with $a_1 \geqslant a_2 \geqslant \ldots \geqslant a_m > 0$, $b_1 \geqslant b_2 \geqslant \ldots \geqslant b_n > 0$. We further suppose that $\mathfrak{A}$ is non-empty. Such a class is called normalized [32]. The assumption of monotone row and column sums sometimes entails no loss of generality (e.g., the class existence problem), but at other times is a restriction (e.g., the zero trace problem).

For given A in $\mathfrak{A}$, the trace σ of A may be defined by $\sigma = \sum_{i=1}^{\min(m,n)} a_{ii}$. Ryser has derived simple formulas for the minimal trace $\tilde{\sigma}$ and the maximal trace $\bar{\sigma}$ of all matrices in $\mathfrak{A}$ [32]. These formulas are in terms of the structure matrix for the class:

(12.22) $$\tilde{\sigma} = \max_{k,l} [\min (k, l) - t_{kl}],$$

(12.23) $$\bar{\sigma} = \min_{k,l} [\max (k, l) + t_{kl}],$$

$$k = 0, 1, \ldots, m; l = 0, 1, \ldots, n.$$

Formulas (12.22) and (12.23) can also be obtained using network flows, but a flow approach would require some of the theory to be developed in

Chapter III. The formula (12.22) for $\bar{\sigma}$ includes Theorem 12.4 as a special case. It is an interesting exercise to demonstrate this directly.

Another problem for the class $\mathfrak{A}$ that has been solved by Ryser is that of determining the maximal term rank $\bar{\rho}$ for all matrices in $\mathfrak{A}$ [31]. Here the term rank ρ of a (0, 1)-matrix is the maximal number of independent 1's in the matrix, or, equivalently, the minimal number of lines that cover all 1's. Ryser's remarkable formula for maximal term rank $\bar{\rho}$ is again in terms of the structure matrix:

$$(12.24) \qquad \bar{\rho} = \min_{k,l} [k + l + t_{kl}],$$

$$k = 0, 1, \ldots, m; l = 0, 1, \ldots, n.$$

No similar formula for minimal term rank $\tilde{\rho}$ is known as of this writing, but Haber has given an effective algorithm for constructing a matrix of term rank $\tilde{\rho}$ [13]. Neither term rank problem appears amenable to a flow approach.

Recently the notion of the "width" ε of a (0, 1)-matrix has been introduced, and a simple formula for the minimal width $\tilde{\varepsilon}$ of matrices in $\mathfrak{A}$ has been found [10]. Here the width of a (0, 1)-matrix A is the least number of columns of A having the property that every row of the resulting submatrix contains at least one 1. That is, viewing A as the incidence matrix of sets *versus* elements, the width of A is the least number of elements that represent all sets. The formula for $\tilde{\varepsilon}$ has a somewhat different character than (12.22), (12.23), and (12.24), but may again be regarded as involving the structure matrix. If we define

$$(12.25) \qquad N(\varepsilon, k, l) = t_{kl} + (s_1 + \ldots + s_\varepsilon) - k\varepsilon,$$

where

$$(12.26) \qquad 0 \leqslant \varepsilon \leqslant n, \qquad 0 \leqslant k \leqslant m, \qquad \varepsilon \leqslant l \leqslant n,$$

then the minimal width $\tilde{\varepsilon}$ for A in $\mathfrak{A}$ is equal to the first non-negative integer ε such that

$$(12.27) \qquad N(\varepsilon, k, l) \geqslant m - k$$

for all integers k and l satisfying (12.26). While $N(\varepsilon, k, l)$ may be defined in terms of the structure matrix by (12.25), it can also be checked that if A in $\mathfrak{A}$ is partitioned thus:

$$A = \begin{bmatrix} * & A_2 & * \\ A_1 & * & A_3 \end{bmatrix},$$

with A_1 of size $m - k$ by ε and A_2 of size k by $l - \varepsilon$, then $N(\varepsilon, k, l)$ is equal to the number of 1's in A_1 plus the number of 0's in A_2 plus the number of 1's in A_3.

The corresponding problem of determining the maximal width $\bar{\varepsilon}$ for all A in $\mathfrak{A}$ appears very difficult.†

The formula (12.27) for $\bar{\varepsilon}$ can be derived using network flows. We sketch this approach. It may first be shown that there is a matrix A in $\mathfrak{A}$ of width $\bar{\varepsilon}$ such that the submatrix composed of the first $\bar{\varepsilon}$ columns of A has at least one 1 in each row. This follows from the monotonicity of the column sums of A and an interchange argument. Here an interchange is a transformation of the elements of A that changes a minor of type

$$\begin{bmatrix} 1 & 0 \\ 0 & 1 \end{bmatrix}$$

into a minor of type

$$\begin{bmatrix} 0 & 1 \\ 1 & 0 \end{bmatrix},$$

or *vice versa*, and leaves all other elements of A fixed.‡

It follows from this observation that $\bar{\varepsilon}$ is the first ε such that the constraints

$$(12.28) \qquad \sum_j a_{ij} = a_i, \qquad \sum_i a_{ij} = b_j,$$

$$(12.29) \qquad \sum_{j=1}^{\varepsilon} a_{ij} \geqslant 1,$$

$$(12.30) \qquad a_{ij} = 0 \quad \text{or} \quad 1,$$

are feasible. Now, using the device of I.11 for bounding partial sums of arc flows emanating from a node, a flow feasibility problem can be set up for the constraints (12.28), (12.29), (12.30). The result is a network of the kind shown in Fig. 12.1 (for $m = 3$, $n = 4$, $\varepsilon = 2$). Here the supplies are a_1, a_2, a_3, the demands b_1, b_2, b_3, b_4, and the arc capacities are unity except for those marked otherwise. Application of the supply-demand theorem to a network having this structure leads, after simplification, to the conditions (12.27) as necessary and sufficient for the existence of the required flow.

It can also be shown that if the $(0, 1)$-matrix rule of this section is applied by first assigning the 1's in the last column, then the next-to-last, and so on, the resulting matrix has minimal width $\bar{\varepsilon}$.

† Especially since a solution to this problem would settle the existence question for finite projective planes. See [33].

‡ The Ryser interchange theorem asserts that if A and A' are in $\mathfrak{A}$, then A is transformable into A' by a finite sequence of interchanges [30].

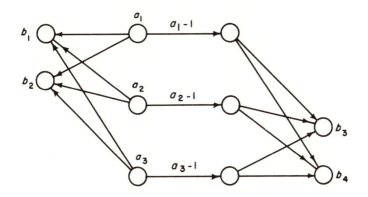

Figure 12.1

References

1. G. B. Dantzig and D. R. Fulkerson, "Minimizing the Number of Tankers To Meet a Fixed Schedule," *Naval Res. Logist. Quart.* 1 (1954), 217–222.
2. G. B. Dantzig and A. J. Hoffman, "Dilworth's Theorem on Partially Ordered Sets," *Linear Inequalities and Related Systems,* Annals of Mathematics Study 38, Princeton University Press, 1956, 207–214.
3. R. P. Dilworth, "A Decomposition Theorem for Partially Ordered Sets," *Ann. of Math.* 51 (1950), 161–166.
4. A. L. Dulmage and N. S. Mendelsohn, "The Term and Stochastic Ranks of a Matrix," *Canad. J. Math.* 11 (1959), 269–279.
5. J. Egerváry, "Matrixok kombinatorikus tulajonságairól," *Mat. és Fiz. Lapok* 38 (1931), 16–28. Translation by H. W. Kuhn, "On Combinatorial Properties of Matrices," *George Washington University Logistics Papers* 11 (1955).
6. L. R. Ford, Jr., and D. R. Fulkerson, "Network Flow and Systems of Representatives," *Canad. J. Math.* 10 (1958), 78–85.
7. D. R. Fulkerson, "Note on Dilworth's Decomposition Theorem for Partially Ordered Sets," *Proc. Amer. Math. Soc.* 7 (1956), 701–702.
8. ———, "A Network Flow Feasibility Theorem and Combinatorial Applications," *Canad. J. Math.* 11 (1959), 440–451.
9. ———, "Zero-one Matrices with Zero Trace," *Pacific J. Math.* 10 (1960), 831–836.
10. ——— and H. J. Ryser, "Widths and Heights of (0, 1)-Matrices," *Canad. J. Math.* 13 (1961), 239–255.
11. D. Gale, "A Theorem on Flows in Networks," *Pacific J. Math.* 7 (1957), 1073–1082.
12. O. Gross, "The Bottleneck Assignment Problem," The RAND Corporation, Paper P-1630, March 6, 1959, presented at the RAND Symposium on Mathematical Programming (Linear Programming and Extensions), March 16–20, 1959.
13. R. M. Haber, "Term Rank of 0, 1 Matrices," *Rend. Sem. Mat. Univ. Padova* 30 (1960), 24–51.

II. FEASIBILITY THEOREMS AND COMBINATORICS

14. M. Hall, Jr., "An Existence Theorem for Latin Squares," *Bull. Amer. Math. Soc.* 51 (1945), 387–388.
15. M. Hall, Jr., "Distinct Representatives of Subsets," *Bull. Amer. Math. Soc.* 54 (1948), 922–926.
16. P. Hall, "On Representatives of Subsets," *J. Lond. Math. Soc.* 10 (1935), 26–30.
17. A. J. Hoffman, "Some Recent Applications of the Theory of Linear Inequalities to Extremal Combinatorial Analysis," *Proc. Symposia on Applied Math.* 10 (1960).
18. —— and H. W. Kuhn, "On Systems of Distinct Representatives," *Linear Inequalities and Related Systems*, Annals of Mathematics Study 38, Princeton University Press, 1956, 199–206.
19. ——, "Systems of Distinct Representatives and Linear Programming," *Amer. Math. Monthly* 63 (1956), 455–460.
20. S. Johnson, "Optimal Two- and Three-stage Production Schedules with Setup Times Included," *Naval Res. Logist. Quart.* 1 (1954), 61–68.
21. D. König, *Theorie der Endlichen und Unendlichen Graphen*, Chelsea Publishing Co., New York, 1950, 258 pp.
22. H. W. Kuhn, "The Hungarian Method for the Assignment Problem," *Naval Res. Logist. Quart.* 2 (1955), 83–97.
23. H. B. Mann and H. J. Ryser, "Systems of Distinct Representatives," *Amer. Math. Monthly* 60 (1953), 397–401.
24. N. S. Mendelsohn and A. L. Dulmage, "Some Generalizations of the Problem of Distinct Representatives," *Canad. J. Math.* 10 (1958), 230–241.
25. O. Ore, "Studies on Directed Graphs I," *Ann. of Math.* 63 (1956), 383–406.
26. ——, "Graphs and Subgraphs," *Trans. Amer. Math. Soc.* 84 (1957), 109–137.
27. R. Rado, "Factorization of Even Graphs," *Quart. J. Math.* 20 (1949), 95–104.
28. J. T. Robacker, "On Network Theory," The RAND Corporation, Research Memorandum RM-1498, May 26, 1955.
29. H. J. Ryser, "A Combinatorial Theorem with an Application to Latin Rectangles," *Proc. Amer. Math. Soc.* 2 (1951), 550–552.
30. ——, "Combinatorial Properties of Matrices of Zeros and Ones," *Canad. J. Math.* 9 (1957), 371–377.
31. ——, "The Term Rank of a Matrix," *Canad. J. Math.* 10 (1958), 57–65.
32. ——, "Traces of Matrices of Zeros and Ones," *Canad. J. Math.* 12 (1960), 463–476.
33. ——, "Matrices of Zeros and Ones," *Bull. Amer. Math. Soc.* 66 (1960), 442–464.
34. W. T. Tutte, "The Factors of Graphs," *Canad. J. Math.* 4 (1952), 314–329.

CHAPTER III

MINIMAL COST FLOW PROBLEMS

Introduction

In this chapter we take up the problem of constructing network flows that minimize cost. The practical importance of this problem area is affirmed by the fact that a sizeable fraction of the linear programming literature has been devoted to it, and an even larger share of the many concrete industrial and military applications of linear programming have been in this domain. Indeed, a survey made in 1956 [65] indicated that about half of such applications at that time fell in the category of transportation problems, or, in our parlance, minimal cost flow problems. Among the reasons for what might seem to be a surprising concentration, particularly in applications, on problems of this kind, are perhaps these: answers to large transportation problems, involving many hundreds of constraints and thousands of variables, can be easily computed, whereas it is an impossible task, at present, to solve a general linear programming problem of these dimensions; a number of linear programs that might not appear to be transportation problems, turn out to be such on closer examination. Well-known examples of this are the warehousing and caterer problems, discussed in this chapter.

The standard transportation problem is that sometimes referred to as the Hitchcock problem, after one of its formulators [42]. Hitchcock also gave a procedure, much akin to the general simplex method, for obtaining a solution. Independently, during World War II, Koopmans arrived at the same problem in connection with his work as a member of the Combined Shipping Adjustment Board. He and Reiter [54] discussed the problem from the standpoint of economic efficiency analysis and pointed out the analogy between it and the classical Maxwell-Kirchhoff electrical network problem. Other writers who studied transportation problems at about the same time are Tolstoi [66], Kantorovitch [50], and Kantorovich and Gavurin [51].

In a paper published in 1951, Dantzig showed how his simplex method for general linear programs specializes to yield an efficient computation for the standard transportation problem [7]. At the same time he observed the important integrity property of solutions to such problems. Since then, various other accounts of the simplex computation for transportation problems have appeared, and other kinds of algorithms have been proposed.

The algorithm described in § 1 for obtaining solutions to the Hitchcock problem is a generalization of a combinatorial procedure developed by Kuhn [56] for the optimal assignment problem (§ 2), a special case of the Hitchcock problem. One of the basic ideas underlying Kuhn's method stems from Egerváry's proof [17] of the König-Egerváry theorem concerning bipartite graphs; another may be regarded as implicit in the proof of this theorem that appears in [55]. In this method, which Kuhn has dubbed the "Hungarian Method," two routines are involved, one for finding a maximal set of independent admissible cells and a minimal covering of admissible cells in an n by n matrix; the other, for transforming to a new set of admissible cells in case the old maximal set contained fewer than n members. For transportation problems, the analogue of the former of these becomes a maximal flow problem; of the latter, a transformation of dual variables. Thus we view the process as one of solving a sequence of maximal flow problems.

We have chosen to discuss the Hitchcock problem, for which the underlying network is bipartite, before presenting a general algorithm for solving minimal cost flow problems in arbitrary networks with capacity constraints on arcs (§ 3). An equivalence between these two problems is then presented in § 4. In § 5 an algorithm is described for finding a shortest chain from one node to another in an arbitrary network in which each arc has an associated length. This problem is a special case of the minimal cost flow problem. In § 6 we return to the latter, allowing arc costs to be negative, but subject to a non-negative directed cycle condition. The shortest chain algorithm of § 5 can be used to initiate the computation in this case. The following two sections (7 and 8) contain brief discussions of the warehousing and caterer problems.

Section 9 applies the theory developed for minimal cost flows to the problem of constructing a maximal dynamic flow for a given number of time periods in a network in which each arc has not only a flow capacity, but a transit time as well. The assumption that capacities and transit times are independent of time leads to a remarkably simple and effective method of solving the maximal dynamic flow problem for all time periods. Without this assumption, the problem can be treated as a static problem in a time-expanded replica of the given network.

Another application of minimal cost flows is discussed in § 10. The problem here is that of determining the least cost for a project composed of many individual jobs that are partially ordered due to technological restrictions. It is assumed that the cost of doing any job varies linearly between given extreme completion times for the job, and a schedule is sought that minimizes project cost, assuming that the project must be completed by a given date.

Section 11 concludes with a description of a method for constructing minimal cost feasible circulations in a network having lower bounds and capacities on arcs. The algorithm of § 11 may be viewed as a generalization of the ones presented earlier in the chapter.

1. The Hitchcock problem

A paraphrase, in the language we have been using, of Hitchcock's statement of the problem might run as follows. Suppose there are m sources $x_1, \ldots, x_m$ for a commodity, with $a(x_i)$ units of supply at x_i, and n sinks $y_1, \ldots, y_n$ for the commodity, with a demand $b(y_j)$ at y_j. If $a(x_i, y_j)$ is the unit cost of shipment from x_i to y_j, find a flow that satisfies demands from supplies and minimizes flow cost.

Since the network for this problem is bipartite, it is convenient to drop the x's and y's and use matrix notation and terminology. Thus, letting $a_i \geqslant 0$ denote the supply at the ith source, $b_j \geqslant 0$ the demand at the jth sink, $a_{ij} \geqslant 0$ the unit shipping cost from source i to sink j, the problem is to find an m by n non-negative matrix f_{ij} that satisfies the row sum constraints

(1.1)
$$\sum_{j=1}^{n} f_{ij} \leqslant a_i, \qquad i = 1, \ldots, m,$$

the column sum constraints

(1.2)
$$\sum_{i=1}^{m} f_{ij} \geqslant b_j, \qquad j = 1, \ldots, n,$$

and minimizes

(1.3)
$$\sum_{i,j} a_{ij} f_{ij}.$$

Since we are assuming $a_{ij} \geqslant 0$, it suffices to suppose that (1.2) are equalities. Indeed, the problem is usually stated with both (1.1) and (1.2) as equalities, and the feasibility assumption that total supply equals total demand, $\sum_i a_i = \sum_j b_j$. However, we shall merely assume that $\sum_i a_i \geqslant \sum_j b_j$, and leave the problem in inequality form.† It follows from the supply-demand theorem that this is a necessary and sufficient condition for feasibility. We shall also assume that the a_i, b_j, and a_{ij} are integers (equivalently, we could suppose they are rationals).

---

† The customary assumption that (1.1) and (1.2) are equalities entails no loss of generality, since one can add an $(n + 1)^{\text{st}}$ column with column sum $\sum_{i=1}^{m} a_i - \sum_{j=1}^{n} b_j$ and take $a_{i,n+1} = 0$, thereby obtaining an equivalent Hitchcock problem in equality form. In order for this simple equivalence to work, it is essential that costs be non-negative.

EXAMPLE. Suppose that unit shipping costs are given by the array of Fig. 1.1 with supplies a_i and demands b_j as indicated in Fig. 1.2. Then a feasible solution with total cost 93 is shown in Fig. 1.3.

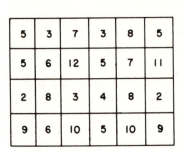

Figure 1.1

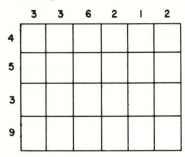

Figure 1.2

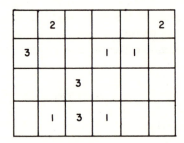

Figure 1.3

To write down the linear programming dual (see I.12) of the Hitchcock problem, we may first rewrite (1.1) as

$$-\sum_j f_{ij} \geqslant -a_i$$

to get the problem in the form (I.12.4)–(I.12.6). Thus, assigning dual "row variables" α_i to the constraints (1.1), dual "column variables" β_j to the constraints (1.2), the dual linear program† has constraints

(1.4) $$-\alpha_i + \beta_j \leqslant a_{ij},$$

(1.5) $$\alpha_i, \beta_j \geqslant 0,$$

subject to which the form

† The dual variables α_i, β_j play a role analogous to potentials in electrical network theory, and are sometimes referred to by this name. They are also frequently given the economic interpretation of prices; this is perhaps more appropriate here, since the primal problem has been verbally described in terms of supplies, demands, and shipping costs. However, we avoid either of these interpretations and simply call them "dual variables," or, later on, "node numbers."

96

(1.6)
$$-\sum_i a_i\alpha_i + \sum_j b_j\beta_j$$

is to be maximized. (If we had taken (1.2) as equalities, the β_j would not be restricted in sign.)

EXAMPLE (continued). A feasible dual solution is shown in Fig. 1.4, where the circled entries indicate equality in the dual constraints (1.4).

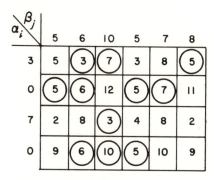

Figure 1.4

Since the form (1.6) has value $-33 + 126 = 93$ for this feasible dual solution, and since (1.6) is bounded above by (1.3) for any pair of feasible solutions to dual and primal, it follows that the feasible solutions shown in Fig. 1.3 and Fig. 1.4 are optimal in their respective problems. Notice that

(a) $-\alpha_i + \beta_j < a_{ij}$ implies $f_{ij} = 0,$

(b) $\alpha_i > 0$ implies $\sum_j f_{ij} = a_i,$

(c) $\beta_j > 0$ implies $\sum_i f_{ij} = b_j.$

Indeed, we know these implications must hold because the f_{ij} and α_i, β_j constitute a pair of optimal solutions to primal and dual.

The general scheme of the algorithm to be described for solving Hitchcock problems is this: start with a particular feasible dual solution α_i, β_j, and attempt to satisfy the primal constraints, allowing positive f_{ij} only if $-\alpha_i + \beta_j = a_{ij}$; more precisely, solve the problem

(1.7)
$$\sum_j f_{ij} \leqslant a_i,$$

(1.8)
$$\sum_i f_{ij} \leqslant b_j,$$

(1.9)
$$f_{ij} \geqslant 0,$$

(1.10)
$$f_{ij} = 0, \qquad\qquad \text{if } -\alpha_i + \beta_j < a_{ij},$$

(1.11)
$$\text{maximize} \sum_{i,j} f_{ij}.$$

This is a maximal flow problem and can consequently be solved by the labeling process. We are then either done, or can use the results of the labeling to transform the old feasible dual solution to an improved one, that is, one that gives a higher value to the dual form (1.6), and a new maximal flow problem emerges. Eventually the computation terminates with optimal solutions to both primal and dual.[†]

It will also be true, although we have not mentioned it explicitly in this sketch of the computation, that, at any stage, $\sum_j f_{ij}$ will equal a_i if the current dual variable α_i is positive. Termination of the process occurs when the demands b_j are met exactly, and thus at termination, the remaining optimality property (c) above will hold.

We proceed to give a detailed statement of the algorithm [21, 22]. To start,[‡] take

$$(1.12) \qquad \alpha_i = 0, \qquad \beta_j = \min_i a_{ij}, \qquad f_{ij} = 0.$$

(Thus the initial α_i, β_j satisfy the dual constraints (1.4), (1.5).) The computation now progresses by a sequence of labelings (A, below); if a labeling results in breakthrough, the flow f is changed appropriately (B, below); if non-breakthrough occurs, the current dual variables α_i, β_j are transformed (C, below).

Cells ij for which $-\alpha_i + \beta_j = a_{ij}$ will be called *admissible*, others *inadmissible*, in describing the computation. The labeling rules are little different from those of II.5. Again we alternate between row and column scanning. This time, however, the flow change will not always be unity. Hence we shall carry along enough information in the labels so that, if breakthrough occurs, the resulting flow change can be effected without first backtracking along the flow augmenting path. (Actually, for hand computation, it is just as convenient to backtrack.)

Routine A (labeling process). Begin by assigning labels $(-, \varepsilon_i)$, where $\varepsilon_i = a_i - \sum_j f_{ij}$, to all rows i for which $\sum_j f_{ij} < a_i$. Next select a labeled row, say row i, and scan it for all (unlabeled) columns j such that cell ij is admissible; label these columns (i, δ_j), where $\delta_j = \varepsilon_i$. Repeat until the labeled rows have all been scanned. Then select a labeled column, say column j, and scan it for all unlabeled rows i such that $f_{ij} > 0$; label such i with (j, ε_i) where $\varepsilon_i = \min (f_{ij}, \delta_j)$. Repeat until previously labeled columns have all been scanned. Then revert to row scanning of newly labeled rows, and so on. If a column is labeled for which $\sum_i f_{ij} < b_j$ (breakthrough), stop the labeling process and apply *Routine* B. Otherwise

[†] A generalization of this primal-dual method to arbitrary linear programs can be found in [11].

[‡] We could start with all $\alpha_i = 0$, all $\beta_j = 0$. However, the starting dual solution (1.12) is a better one.

continue until no more labels can be assigned (non-breakthrough) and go to *Routine* C.

Routine B (flow change). Here we have labeled column j with (i, δ_j) and $\sum_i f_{ij} < b_j$. Alternately add and subtract $\varepsilon = \min(\delta_j, b_j - \sum_i f_{ij})$ along the path indicated by the labels. That is, add ε to f_{ij}, then proceed, in row i, to the cell singled out by the first member of the label on row i, and subtract ε, then proceed, in the column reached, to the position picked out by the first member of its label, add ε, and so on, stopping when one of the initially labeled rows has been reached. If now all column demands have been satisfied, the algorithm terminates. Otherwise, start with the new flow generated, discard the old labels, and go back to *Routine* A.

Routine C (dual variable change). The labeling process has resulted in non-breakthrough. Let I and J be the index sets of labeled rows and columns, respectively, and define new dual variables by

$$(1.13) \qquad \alpha_i' = \begin{cases} \alpha_i, & i \in I, \\ \alpha_i + \delta, & i \in \bar{I}, \end{cases}$$

$$(1.14) \qquad \beta_j' = \begin{cases} \beta_j, & j \in J, \\ \beta_j + \delta, & j \in \bar{J}, \end{cases}$$

where

$$(1.15) \qquad \delta = \min_{I\bar{J}} (a_{ij} + \alpha_i - \beta_j).$$

The labeling process is then repeated with new admissible cells defined by α_i', β_j'.

Before showing that this algorithm solves the Hitchcock problem in a finite number of steps, we make some preliminary observations dealing with the non-breakthrough case. First of all, the dual variable change δ of *Routine* C is a positive integer, since all cells ij in the (non-empty) rectangle $I\bar{J}$ are inadmissible relative to α_i and β_j; for, otherwise, some column of $\bar{J}$ would have been labeled from some row of I. (That the rectangle $I\bar{J}$ is non-empty follows from the assumption that termination has not yet occurred, for if either I or $\bar{J}$ were empty, the minimal covering value $a(\bar{I}) + b(J)$ for all admissible cells would be at least $\sum_{j=1}^n b_j$, which is absurd since $a(\bar{I}) + b(J) = \sum_{i,j} f_{ij}$, the flow value.) Thus we gain new admissible cells in $I\bar{J}$ corresponding to positions for which the minimum in (1.15) is attained. On the other hand, any admissible cell in $\bar{I}J$ becomes inadmissible for the next labeling. For such a cell, however, we must have had $f_{ij} = 0$, as otherwise a row of $\bar{I}$ would have been labeled from a column of J, and thus the old flow can be used to start the next labeling. Indeed, even the old labeling can be retained to initiate the new one, because

(a) the pattern of admissibility in IJ and $\bar{I}\bar{J}$ has not been altered, and

(b) the admissible cells lost in $\bar{I}J$ could not have contributed to the old labeling, since J received labels from I.

That the algorithm terminates in a finite number of steps can be seen as follows. First of all, since each occurrence of breakthrough increases the flow value $\sum_{i,j} f_{ij}$ by at least one unit, the number of labelings that produce breakthrough is bounded above by the total demand $\sum_j b_j$. On the other hand, one can show easily that each non-breakthrough increases the dual form (1.6) by at least one unit. Since this form is bounded above (e.g., by $\sum_{i,j} a_{ij} f_{ij}$ for any feasible f), this will establish finiteness. To see that the dual form increases, note from (1.13) and (1.14) that

$$-\sum_i a_i \alpha_i' + \sum_j b_j \beta_j' = -\sum_i a_i \alpha_i + \sum_j b_j \beta_j + \delta(b(\bar{J}) - a(\bar{I})).$$

As we have observed, δ is a positive integer. Thus it suffices to show that $b(\bar{J}) - a(\bar{I}) > 0$. Again this strict inequality follows from the fact that if non-breakthrough has occurred, then

$$a(\bar{I}) + b(J) = \sum_{i,j} f_{ij} < \sum_{j=1}^n b_j.$$

Another way to demonstrate finiteness rests on the fact that the number of consecutive occurrences of non-breakthrough can never exceed $\min(m, n) + 1$. For, as was pointed out, after non-breakthrough, the old labeling can be repeated. In addition, at least one more row and column can be labeled. To see this, recall that after changing the dual variables, at least one new admissible cell is obtained in the rectangle $I\bar{J}$. Thus some column of $\bar{J}$ will receive a label. If the new labeling is to result again in non-breakthrough, then this column demand has been fulfilled; that is, $\sum_i f_{ij} = b_j$ for the column $\bar{j} \in \bar{J}$ in question. Since $f_{ij} = 0$ for $i \in I$, then $f_{ij} > 0$ for some $i \in \bar{I}$, and consequently another row can be labeled. (We are tacitly assuming that columns with zero demand, if any, have been deleted from the problem to start with.)

This second finiteness proof leads to a bound on the total number of labelings required to solve a Hitchcock problem that depends only on the total demand and the size of the problem, the bound being

(1.16)
$$\sum_{j=1}^n b_j + \left(\sum_{j=1}^n b_j - 1\right)(\min(m, n) + 1).$$

Here the first term bounds the number of labelings resulting in breakthrough; the second, the number of labelings resulting in non-breakthrough. While this bound, the idea for which is due to Munkres [61], does not come close to being achieved in practice, it is still sufficiently good to be interesting. It perhaps also lends some theoretical support to the empirical observation that "long, narrow" Hitchcock problems are faster solved than "comparable square" ones, the comparison being on $m + n$, the number of constraints.

100

Having established finiteness, our next job is to show that, upon termination of the computation, a pair of optimal solutions to primal and dual has been constructed. This is almost obvious, but nonetheless we give the details. First note that it is clear from (1.13)–(1.15) that feasible dual solutions are maintained throughout the computation, provided we start with one. It is also clear that, upon termination, a feasible primal solution has been constructed. Moreover, at each stage we allow the possibility $f_{ij} > 0$ only if the current dual variables satisfy $a_{ij} + \alpha_i - \beta_j = 0$, and thus upon termination we have this optimality property. Also, at termination we have $\sum_i f_{ij} = b_j$ for all j. Consequently, the optimality criterion that $\beta_j > 0$ should imply $\sum_i f_{ij} = b_j$ is satisfied automatically. The third and last optimality condition that needs to be checked is that, at termination, $\alpha_i > 0$ only if $\sum_j f_{ij} = a_i$. This certainly holds for the starting α_i. It also holds for all subsequent α_i because α_i increases from zero only if $i \in I$ at some stage. This means that at this stage $\sum_j f_{ij} = a_i$, and hence this equality is maintained for later stages.

We summarize some of this discussion in the following theorem.

THEOREM 1.1. *Let* $a_i > 0$ $(i = 1, \ldots, m)$, $b_j > 0$ $(j = 1, \ldots, n)$, *and* $a_{ij} \geqslant 0$ *be given integers. The algorithm composed of Routines* A, B, C *produces (integral) solutions to the corresponding Hitchcock problem and its dual after at most* $\sum_j b_j + (\sum_j b_j - 1)$ $(\min (m, n) + 1)$ *applications of Routine* A.

In computing a Hitchcock problem by hand, it is convenient to carry along two arrays, one of which might be termed the "cost-dual variable array" and the other, the "flow array." In the first of these the dual variables are recorded above and to the left, say, of the cost array; using the resulting array, it is easy to locate cells for which $a_{ij} + \alpha_i - \beta_j = 0$. These can be marked by circles in the flow array. The labeling process is then carried out on the flow array, labels being recorded to the right and below the array, say. If breakthrough results, the indicated changes are made in the flow array and the old labels erased. If non-breakthrough results, the dual variables are changed in the cost-dual variable array, new admissible cells are marked by circles in the flow array, and the circles in old admissible cells that are no longer admissible are erased. One can then re-label, using the old flow and old labeling to initiate the new labeling.

It is also convenient to carry along an extra row and column in the flow array for the purpose of recording the remaining supplies $a_i - \sum_j f_{ij}$ and demands $b_j - \sum_i f_{ij}$, changing these with each flow change. Those $b_j - \sum_i f_{ij}$ that are positive indicate potential breakthrough columns; the positive $a_i - \sum_j f_{ij}$ single out rows that start the labeling process.

EXAMPLE (continued). The figures and interspersed comment below

101

describe the step-by-step solution of the example. In carrying out the labeling process, we have not recorded the second members of the labels, but have found the flow change by backtracking in the event of break-through.

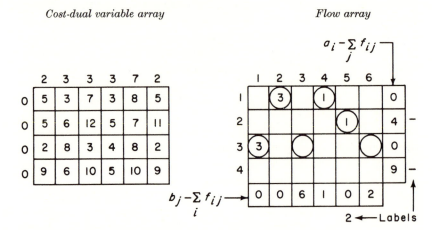

Cost-dual variable array   *Flow array*

The initial flow was filled in by sweeping the first row for its first circle, here cell 12, setting $f_{12} = \min(a_1, b_2) = 3$, reducing the supply and demand for row 1 and column 2 by 3, then going on to the next circle, and so on. (Of course we could have repeated the labeling process a number of times instead.) We then label and do not break through to one of the "shorted" columns 3, 4, 6. Thus $I = \{2, 4\}$, $\bar{J} = \{1, 2, 3, 4, 6\}$ and we turn to the cost-dual variable array to compute $\delta = \min_{I\bar{J}} (a_{ij} + \alpha_i - \beta_j) = 2$, the minimum being achieved in cells 24 and 44. We then change the dual variables using (1.13), (1.14), add circles in cells 24, 44, and remove circles in $\bar{I}J$ (here there are none), to obtain the new arrays shown below.

	4	5	5	5	7	4
2	5	3	7	3	8	5
0	5	6	12	5	7	11
2	2	8	3	4	8	2
0	9	6	10	5	10	9

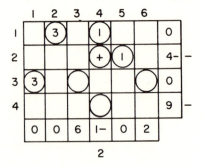

102

The labeling has resulted in breakthrough to column 4, the resulting flow change of $\varepsilon = 1$ being indicated by $+$ and $-$ in the flow array.

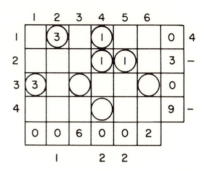

Non-breakthrough, $\delta = 1$, new circle in cell 21, no circles lost.

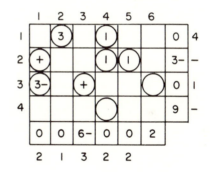

Breakthrough to column 3, $\varepsilon = 3$.

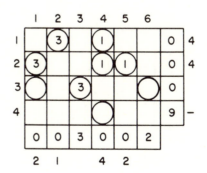

103

Non-breakthrough, $\delta = 2$, new circle 16, lose circle 31.

	5	5	8	5	7	7
2	5	3	7	3	8	5
0	5	6	12	5	7	11
5	2	8	3	4	8	2
0	9	6	10	5	10	9

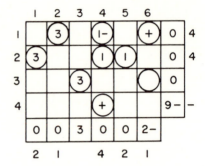

Breakthrough to column 6, $\varepsilon = 1$.

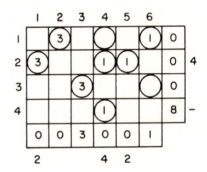

Non-breakthrough, $\delta = 1$, new circles 22, 42, lose circle 14.

	5	6	9	5	7	8
3	5	3	7	3	8	5
0	5	6	12	5	7	11
6	2	8	3	4	8	2
0	9	6	10	5	10	9

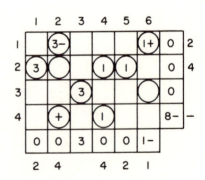

104

Breakthrough to column 6, $\varepsilon = 1$.

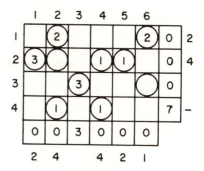

Non-breakthrough, $\delta = 1$, new circles 13, 43, lose circle 36.

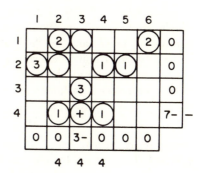

Breakthrough to column 3, $\varepsilon = 3$, and problem done.

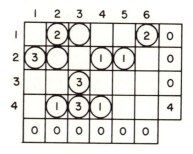

The last cost-dual variable array gives an optimal dual solution; the last flow array, an optimal primal solution. These are the same as shown previously in Fig. 1.3 and Fig. 1.4.

If desired, alternative minimizing flows, if such exist, can be found by using the optimal dual solution. The additional constraints that must be

105

observed are just the optimality properties previously stated. For instance, we can permit f_{13} or f_{22} to be positive, since they correspond to admissible cells in the last flow array, so long as we insist that $f_{ij} = 0$ for inadmissible cells and $\sum_j f_{1j} = a_1$, $\sum_j f_{3j} = a_3$ (since $\alpha_1 > 0$, $\alpha_3 > 0$). Also, of course, we must maintain the conditions $\sum_i f_{ij} = b_j$ (since all $\beta_j > 0$). Thus, for example, to look for a minimizing flow in which $f_{13} > 0$, we could impose a lower bound of unity on f_{13} and solve the resulting flow problem on the "network" of admissible cells.

In a similar way, the optimal dual variables can be used to discover how much the shipping bill would increase if a positive flow were allowed in an inadmissible cell. For instance, increasing f_{23} from 0 to k would increase the shipping bill by $k(a_{23} + \alpha_2 - \beta_3) = 2k$, so long as flows corresponding to other inadmissible cells were held at zero and the other optimality conditions preserved. Consequently the "most expensive" route to use is 32, for which $a_{32} + \alpha_3 - \beta_2 = 9$ is maximal.

Other information is contained in an optimal dual solution. For example, since $\alpha_3 = 7$, it follows that increasing the supply at source 3 by one unit is potentially worth 7 units on the transportation bill (here it really is worth 7 units), and so on. Linear programming facts of this kind are well known and we shall not dwell on them, except to say that for the Hitchcock problem such statements follow readily from the identity

$$\sum_{i,j} a_{ij}(f'_{ij} - f_{ij}) = \sum_{i,j} (a_{ij} + \alpha_i - \beta_j)f'_{ij} - \sum_i \alpha_i(a'_i - a_i) + \sum_j \beta_j(b'_j - b_j).$$

Here f is an optimal primal solution, α, β, an optimal dual solution, and f' a flow from sources to sinks having row sums a'_i and column sums b'_j.

Another rule for changing the dual variables that can be used instead of the one given by (1.13)–(1.15) has been stated by Flood [19]. This transformation has the advantage of making larger changes in the dual form at the expense of destroying old admissible cells containing positive flow entries.

To deduce this rule, let us begin by supposing that, instead of adding as large a constant as possible to all α_i, $i \in \bar{I}$, and β_j, $j \in \bar{J}$, maintaining dual feasibility, we look, more generally, for a dual variable change of the following kind:

$$(1.17) \qquad \alpha'_i = \begin{cases} \alpha_i, & i \in I, \\ \alpha_i + \delta, & i \in \bar{I}, \end{cases}$$

$$(1.18) \qquad \beta'_j = \begin{cases} \beta_j, & j \in J, \\ \beta_j + \delta_j, & j \in \bar{J}, \end{cases}$$

where δ, $\delta_j \geq 0$ and $a_{ij} + \alpha'_i - \beta'_j \geq 0$, the latter for all i, j. For this transformation, the change in dual form is given by

$$(1.19) \qquad \sum_{\bar{J}} b_j \delta_j - \delta \sum_{\bar{I}} a_i,$$

106

and thus we take as our objective that of maximizing (1.19) subject to the constraints

$$(1.20) \qquad a_{ij} + \alpha_i - \beta_j - \delta_j \geq 0, \qquad\qquad ij \in I\bar{J},$$

$$(1.21) \qquad a_{ij} + \alpha_i - \beta_j + \delta - \delta_j \geq 0, \qquad ij \in \bar{I}\bar{J},$$

$$(1.22) \qquad\qquad \delta \geq 0, \qquad \delta_j \geq 0.$$

The reason for restricting attention to the transformation (1.17), (1.18), instead of allowing variable quantities to be added to the α_i, $i \in I$, as well, is that for the former, the resulting maximum problem can be solved by a simple rule, as we shall see. Without the restriction, we have, in effect, a problem on a par with the original Hitchcock problem.

To simplify the notation, set

$$(1.23) \qquad\qquad \bar{a}_{ij} = a_{ij} + \alpha_i - \beta_j.$$

Notice that on the first step, that is, for the α_i, β_j defined by (1.12), it is true that every column of the rectangle $\bar{I}\bar{J}$ contains at least one $\bar{a}_{ij} = 0$, because for any j, $\bar{a}_{ij} = 0$ for at least one i, and the rectangle $I\bar{J}$ contains no zero $\bar{a}_{ij}$. We may assume that we have this property in the maximum problem (1.19)–(1.22), since we can guarantee it, if necessary, by increasing the β_j until every column contains at least one $\bar{a}_{ij} = 0$.

For fixed δ, the change (1.19) is maximized by taking each δ_j as large as possible, that is,

$$\delta_j = \min\,[\min_I \bar{a}_{ij}, \min_{\bar{I}}\,(\bar{a}_{ij} + \delta)], \qquad\qquad j \in \bar{J}.$$

In view of the remark of the preceding paragraph, this becomes

$$\delta_j = \min\,(\min_I \bar{a}_{ij}, \delta), \qquad\qquad j \in \bar{J}.$$

Then our problem is to determine a δ that achieves

$$(1.24) \qquad\qquad \max_{\delta \geq 0} \sum_{\bar{J}} b_j \min\,(\delta, \lambda_j) - \delta \sum_{\bar{I}} a_i,$$

where

$$(1.25) \qquad\qquad \lambda_j = \min_I \bar{a}_{ij}, \qquad\qquad j \in \bar{J}.$$

Let us assume the notation has been chosen so that $\bar{J} = \{1, \ldots, q\}$, and $0 < \lambda_1 \leq \lambda_2 \leq \ldots \leq \lambda_q$. We first remark that the maximum in (1.24) is attained by selecting δ to be one of the λ_j $(j = 1, \ldots, q)$. Indeed, any δ'

such that $\lambda_k < \delta' < \lambda_{k+1}$ $(k = 1, \ldots, q - 1)$ is majorized by selecting either $\delta = \lambda_k$ or $\delta = \lambda_{k+1}$, since, of the three numbers

$$\sum_{j=1}^{k} b_j\lambda_j + \lambda_k\left[\sum_{j=k+1}^{q} b_j - \sum_{\bar{I}} a_i\right],$$

$$\sum_{j=1}^{k} b_j\lambda_j + \delta'\left[\sum_{j=k+1}^{q} b_j - \sum_{\bar{I}} a_i\right],$$

$$\sum_{j=1}^{k} b_j\lambda_j + \lambda_{k+1}\left[\sum_{j=k+1}^{q} b_j - \sum_{\bar{I}} a_i\right],$$

the first is no smaller than the second if the quantity in parentheses is non-positive, and the third is larger than the second otherwise. Moreover, if $\delta' > \lambda_q$, the comparison is between the two numbers

$$\sum_{j=1}^{q} b_j\lambda_j - \delta'\sum_{\bar{I}} a_i, \qquad \sum_{j=1}^{q} b_j\lambda_j - \lambda_q\sum_{\bar{I}} a_i,$$

the second being larger. If $\delta' < \lambda_1$, the comparison is between the numbers

$$\delta'\left[\sum_{j=1}^{q} b_j - \sum_{\bar{I}} a_i\right], \qquad \lambda_1\left[\sum_{j=1}^{q} b_j - \sum_{\bar{I}} a_i\right],$$

and again the second is larger since the number in parentheses is positive. Thus δ is one of $\lambda_1, \ldots, \lambda_q$, and the problem is reduced to determining

$$\max_{k} \varphi(k) = \max_{k}\left[\sum_{j=1}^{k} b_j\lambda_j + \lambda_k\left(\sum_{j=k+1}^{q} b_j - \sum_{\bar{I}} a_i\right)\right]$$

for $k = 1, \ldots, q$. Now

$$\varphi(k + 1) - \varphi(k) = (\lambda_{k+1} - \lambda_k)\left(\sum_{j=k+1}^{q} b_j - \sum_{\bar{I}} a_i\right),$$

and thus $\varphi(k + 1) - \varphi(k)$ is non-negative or non-positive according as the second factor on the right is. It follows that the maximum problem is solved by selecting $\delta = \lambda_k$, where k is the first integer for which

$$\sum_{j=1}^{k} b_j \geqslant \sum_{j=1}^{q} b_j - \sum_{\bar{I}} a_i.$$

Thus we have at our disposal the following alternative rule for changing dual variables after non-breakthrough.

Maximal dual change rule. Let

(1.26) $$\lambda_j = \min_{I}\,(a_{ij} + \alpha_i - \beta_j), \qquad j \in \bar{J},$$

and arrange the λ_j in increasing order. Beside each λ_j record the corresponding column demand b_j. Accumulate these latter until the sum first exceeds $b(\bar{J}) - a(\bar{I})$, and let δ denote the corresponding λ_j. Then define

$$(1.27) \qquad \alpha_i' = \begin{cases} \alpha_i, & i \in I, \\ \alpha_i + \delta, & i \in \bar{I}, \end{cases}$$

$$(1.28) \qquad \beta_j' = \min_i (a_{ij} + \alpha_i').$$

Notice that taking $\delta = \lambda_1$ corresponds to the simpler transformation (1.13)–(1.15). On the other hand, if the above rule is used, admissible cells in the rectangle $\bar{I}\bar{J}$ may become inadmissible for the next labeling; since these may contain positive entries, the flow value may be temporarily decreased so that more applications of *Routine* A would be required before non-breakthrough again occurs. It seems likely, however, particularly for machine computation, that use of the maximal dual change rule in place of the simpler one would be worth while.†

EXAMPLE (continued). After the initial step in solving this example, the following arrays were obtained:

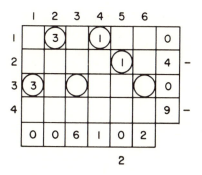

---

† The basic algorithm, using the transformation (1.13)–(1.15), has been coded for the IBM 704 computer by K. Speilberg; this code is available through SHARE, No. 464. Computing times quoted for the code (exclusive of input-output times) for several examples were:

matrix	time
130 by 30	1 min. 30 sec.
160 by 30	3 min. 34 sec.
190 by 30	4 min. 0 sec.
220 by 30	4 min. 58 sec.

Comparisons were also made with the specialized form of the simplex algorithm that is known as the stepping-stone method. For the same set of problems, the latter times were 2 min. 13 sec., 4 min. 56 sec., 7 min. 5 sec., and 11 min. 58 sec., respectively.

Suppose we apply the maximal dual change rule. Here $\tilde{I} = \{1, 3\}$, $\tilde{J} = \{1, 2, 3, 4, 6\}$, $b(\tilde{J}) - a(\tilde{I}) = 13 - 7 = 6$, and

$$\lambda_1 = 3, \quad \lambda_2 = 3, \quad \lambda_3 = 7, \quad \lambda_4 = 2, \quad \lambda_6 = 7,$$
$$b_1 = 3, \quad b_2 = 3, \quad b_3 = 6, \quad b_4 = 2, \quad b_6 = 2.$$

Thus, since $b_4 + b_1 = 5 < 6 < b_4 + b_1 + b_2 = 8$, we take $\delta = \lambda_2 = 3$. Then the new arrays are

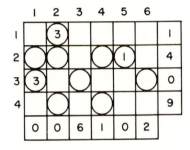

Observe that the flow value has been decreased, but many more admissible cells are available. Three successive breakthroughs are then followed by the non-breakthrough situation

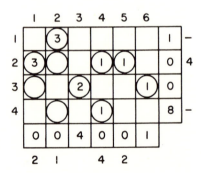

with $\tilde{I} = \{3\}$, $\tilde{J} = \{3, 6\}$, $b(\tilde{J}) - a(\tilde{I}) = 5$, and

$$\lambda_3 = 4, \quad \lambda_6 = 3,$$
$$b_3 = 6, \quad b_6 = 2.$$

110

Thus $\delta = \lambda_3 = 4$, and we obtain the new arrays

	5	6	10	5	7	8
3	5	3	7	3	8	5
0	5	6	12	5	7	11
7	2	8	3	4	8	2
0	9	6	10	5	10	9

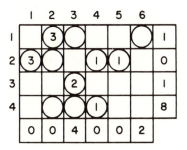

At this point we have an optimal dual solution, and are therefore almost done. Several more labelings (at least three) are required to construct an optimal primal solution.

2. The optimal assignment problem [56, 57, 60, 61, 68, 69]

A well-known extremal combinatorial problem, which generalizes the problem of assigning qualified personnel to jobs (II.5) and is also similar in some respects to the bottleneck assignment problem (II.6), is that of assigning n men to n jobs in an optimal fashion: it is assumed that man i in job j has an efficiency measured by the non-negative integer a_{ij}, and a permutation or assignment $i \to P(i)$ is sought that maximizes the efficiency sum

$$\sum_{i=1}^{n} a_{i,P(i)}.$$

If one takes $a_{ij} = 1$ or 0 according as man i is or is not qualified for job j, the problem of II.5 is obtained.

Since, by Theorem 1.1, a Hitchcock problem with integral supplies and demands always has an integral solution, the optimal assignment problem can be posed as the special Hitchcock problem :†

$$(2.1) \qquad \sum_{j=1}^{n} f_{ij} = 1, \qquad\qquad i = 1, \ldots, n,$$

$$(2.2) \qquad \sum_{i=1}^{n} f_{ij} = 1, \qquad\qquad j = 1, \ldots, n,$$

† That the combinatorial problem can be solved as a linear programming problem can be seen in many ways. For example, as is well known, the permutation matrices are the extreme points of the convex set of doubly stochastic matrices; that is, the convex set defined by (2.1)–(2.3), a fact that is readily deducible, for instance, from Hall's theorem. Thus any linear programming method that constructs an extreme point solution, as the simplex method does, would solve the assignment problem. While our algorithm for Hitchcock problems does not guarantee an extreme point solution, it does guarantee an integral solution, and this suffices.

111

$$(2.3) \qquad\qquad\qquad f_{ij} \geqslant 0,$$

$$(2.4) \qquad\qquad\qquad \text{maximize} \sum_{i,j} a_{ij} f_{ij}.$$

(One could equally well assume m men and n jobs with $m \geqslant n$, and state the problem in mixed equality-inequality form, insisting that every job be filled precisely once, and no man be assigned to more than one job.)

Except for the fact that maximization has replaced minimization, the problem is now cast in familiar form. We can get rid of this slight difference, if desired, by subtracting each a_{ij} from, say, $K = \max_{i,j} a_{ij}$, and then minimizing

$$\sum_{i,j} (K - a_{ij}) f_{ij} = Kn - \sum_{i,j} a_{ij} f_{ij}.$$

In this form the algorithm of the last section requires no restatement to be applicable. Or, one could use the dual of (2.1)–(2.4), which can be written as

$$(2.5) \qquad\qquad -\alpha_i + \beta_j \geqslant a_{ij}, \qquad\qquad i, j = 1, \ldots, n,$$

$$(2.6) \qquad\qquad \text{minimize} \sum_j \beta_j - \sum_i \alpha_i,$$

to make necessary changes in the computation.

Since row and column sums are unity, each maximal flow problem encountered in solving an optimal assignment problem is of König-Egerváry type; that is, a maximal set of independent admissible cells is to be constructed.

It follows from Theorem 1.1 that the number of labelings required to solve an optimal assignment problem can not exceed $n^2 + n - 1$. Actually a somewhat better bound can be obtained as follows. Suppose that a breakthrough has produced a partial assignment containing $v < n$ ones, and this is followed by non-breakthrough with labeled rows I and labeled columns J. Then $|\bar{I}| + |J| = v$ and hence $|\bar{I}| \leqslant v$. Consequently, since $|\bar{I}|$ decreases by at least one with each additional consecutive non-breakthrough, there can be at most $v + 1$ such. It follows that the total number of labelings is no greater than $\frac{1}{2}(n^2 + 3n - 2)$, which is a uniformly better bound than the one given by Theorem 1.1.

While the optimal assignment problem is a special case of the Hitchcock problem, it is equally true that a Hitchcock problem with integral row and column sums a_i and b_j can be solved as an optimal assignment problem. For, taking the Hitchcock problem in equality form, we may replace the ith row by a_i rows, each having unit row sum and all having the same costs a_{ij}; and similarly for the columns, thereby obtaining an equivalent n by n assignment problem with $n = \sum_i a_i = \sum_j b_j$. For computational purposes, expanding the problem in this way is of course not worth while.

3. The general minimal cost flow problem†

The construction of minimal cost flows that fulfill demands at some nodes from supplies at others, in a network in which each arc has infinite capacity and a unit shipping cost, assumed non-negative, has been referred to in the linear programming literature as the trans-shipment problem. Orden [62] and others have recognized the fact that such a problem can be converted to a Hitchcock problem. In the next section, we shall present an equivalence between the trans-shipment problem with capacity constraints on arcs (the minimal cost flow problem) and the Hitchcock problem, but our aim in this section is to treat the former directly.

Assume given a network $[N; \mathscr{A}]$ having sources S, intermediate nodes R, and sinks T, arc capacities $c(x, y)$, arc costs $a(x, y)$, with supplies $a(x)$ for $x \in S$ and demands $b(x)$ for $x \in T$. The problem is to construct a feasible flow, if one exists, which minimizes cost. That is, we wish to solve the linear program

$$(3.1) \qquad f(x, N) - f(N, x) \leqslant a(x), \qquad\qquad x \in S,$$

$$(3.2) \qquad f(x, N) - f(N, x) = 0, \qquad\qquad x \in R,$$

$$(3.3) \qquad f(x, N) - f(N, x) \leqslant -b(x), \qquad\qquad x \in T,$$

$$(3.4) \qquad 0 \leqslant f(x, y) \leqslant c(x, y), \qquad\qquad (x, y) \in \mathscr{A},$$

$$(3.5) \qquad \text{minimize} \sum_{\mathscr{A}} a(x, y) f(x, y).$$

We suppose throughout that $a(x)$, $b(x)$, $c(x, y)$ are positive integers, $a(x, y)$ non-negative integers.

It is convenient, particularly for later discussion of other problems, to put this problem in slightly different terms. The differences are these. First, we shall assume a single source and single sink in our network $[N; \mathscr{A}]$;‡ second, we introduce the flow value v as an explicit variable in the program; third, instead of minimizing the function in (3.5), we shall maximize

$$(3.6) \qquad pv - \sum_{\mathscr{A}} a(x, y) f(x, y).$$

Here p may be thought of as a parameter that consecutively assumes the values $0, 1, 2, \ldots$. Thus we are now considering the sequence of programs

$$(3.7) \qquad f(s, N) - v = 0,$$

$$(3.8) \qquad f(x, N) - f(N, x) = 0, \qquad\qquad x \neq s, t,$$

† The algorithm of this section was originally developed as a means of solving the maximal dynamic flow problem [24]. This problem will be discussed later in the chapter (§9).

‡ We may do this by adding to the original network nodes s, t, together with source and sink arcs (s, S), (T, t), taking $c(s, x) = a(x)$ for $x \in S$, $c(x, t) = b(x)$ for $x \in T$, and $a(s, x) = 0$ for $x \in S$, $a(x, t) = 0$ for $x \in T$.

(3.9) $$-f(N, t) + v = 0,$$

(3.10) $$0 \leqslant f(x, y) \leqslant c(x, y), \qquad\qquad (x, y) \in \mathscr{A},$$

(3.11) $$\text{maximize } pv - \sum_{\mathscr{A}} a(x, y)f(x, y).$$

We term the pth program of this sequence the pth *related program*. The function to be maximized in the pth related program places a premium of p units on every unit of flow that gets through the network, and a penalty $a(x, y)$ on every unit of flow in arc (x, y).

For p sufficiently large, the related program asks for a maximal flow that minimizes cost over all maximal flows. Hence for large p, a solution either solves the original flow problem (3.1)–(3.5) (if v equals the total demand), or shows the latter to be infeasible. (Of course, feasibility could be determined to begin with by solving a maximal flow problem.) The computation will generate successive flows $f_0 = 0, f_1, \ldots, f_P$, the last being a maximal flow. Each f_p will be a solution to the pth related program, and thus P will be a "sufficiently large" value of p.

Before describing this computation, let us motivate it by considering the dual of the pth related program. If we assign dual variables $\pi(x)$ to equations (3.7)–(3.9), and $\gamma(x, y)$ to the capacity constraints $f(x, y) \leqslant c(x, y)$, then the dual has constraints

(3.12) $$-\pi(s) + \pi(t) = p,$$

(3.13) $$\pi(x) - \pi(y) + \gamma(x, y) \geqslant -a(x, y), \qquad (x, y) \in \mathscr{A},$$

(3.14) $$\gamma(x, y) \geqslant 0, \qquad\qquad (x, y) \in \mathscr{A},$$

subject to which the form

(3.15) $$\sum_{\mathscr{A}} c(x, y)\gamma(x, y)$$

is to be minimized. (The equality appears in (3.12) since the sign of v has not explicitly been restricted, although one could do so.) We refer to the dual variables $\pi(x)$ associated with the nodes of the network as *node numbers*,† the dual variables $\gamma(x, y)$ as *arc numbers*. Although the node numbers are not necessarily non-negative, the ones constructed by the algorithm will be, and in fact, non-negative integers satisfying $\pi(s) = 0$, $\pi(t) = p$.

Optimality criteria for primal and dual are

(3.16) $$\pi(x) - \pi(y) + \gamma(x, y) > -a(x, y) \Rightarrow f(x, y) = 0,$$

(3.17) $$\gamma(x, y) > 0 \Rightarrow f(x, y) = c(x, y).$$

---

† Again the reader who is so inclined may wish to interpret these as potentials or prices.

It follows that if a flow f and node numbers π can be constructed satisfying

(3.18) $$\pi(s) = 0, \qquad \pi(t) = p,$$

(3.19) $$\pi(y) - \pi(x) < a(x, y) \Rightarrow f(x, y) = 0,$$

(3.20) $$\pi(y) - \pi(x) > a(x, y) \Rightarrow f(x, y) = c(x, y),$$

then, by taking

(3.21) $$\gamma(x, y) = \max (0, \pi(y) - \pi(x) - a(x, y)),$$

feasible solutions to both the pth related program and its dual have been found that satisfy (3.16)–(3.17), and are therefore optimal in their respective programs. Consequently, we shall make no explicit mention of the arc numbers, but shall deal only with flows and node numbers, aiming at the optimality properties (3.18), (3.19), (3.20).

Although we have used general duality principles in arriving at these optimality properties, it is easy to check directly that these properties imply that f minimizes $\sum a(x, y) f(x, y) - pv$. For if we define

(3.22) $$\bar{a}(x, y) = a(x, y) + \pi(x) - \pi(y),$$

it follows that

$$\sum_{x,y} \bar{a}(x, y) f(x, y) = \sum_{x,y} a(x, y) f(x, y) + \sum_{x,y} \pi(x) f(x, y) - \sum_{x,y} \pi(y) f(x, y)$$

$$= \sum_{x,y} a(x, y) f(x, y) + \sum_x \pi(x) \sum_y f(x, y) - \sum_y \pi(y) \sum_x f(x, y)$$

$$= \sum_{x,y} a(x, y) f(x, y) + \sum_x \pi(x) \sum_y [f(x, y) - f(y, x)]$$

$$= \sum_{x,y} a(x, y) f(x, y) + v[\pi(s) - \pi(t)].$$

Hence, by (3.18),

(3.23) $$\sum \bar{a}(x, y) f(x, y) = \sum a(x, y) f(x, y) - pv.$$

Now (3.19) and (3.20) clearly imply that f minimizes the left-hand side of (3.23), hence also the right.

As was the case for the Hitchcock problem, the minimal cost flow algorithm of this section consists, in essence, of solving a sequence of maximal flow problems, each on the sub-network of admissible arcs, one difference being that now there will be two kinds of inadmissible arcs, corresponding to (3.19) and (3.20). For the former of these, the arc flow will be held fixed at zero, while for the latter, the saturation condition will be maintained. Thus each maximal flow problem can be thought of as one in which upper bounds of zero are imposed on certain arcs, and lower bounds equal to the arc capacity, on others. (Each such maximal flow problem will be feasible.) After solving by the labeling method, the current node numbers are changed by adding a constant to all node numbers corresponding to unlabeled nodes. In the description below, we have assumed this constant to be 1. This is merely a descriptive convenience: it corresponds to changing the parameter p to $p + 1$ instead of some possibly larger value $p + \delta$. We shall state later how large to take δ in actual practice.

There are alternative ways of starting the computation. The simplest is to begin with all node numbers equal to zero and the zero flow. This corresponds to taking $p = 0$, since (3.18), (3.19), (3.20) are then satisfied.

To describe the general cycle of the computation, let us now suppose that we have an integral flow f and node integers π satisfying (3.18), (3.19), (3.20) with $\pi(t) = p$ (that is, f solves the pth related program), and we wish to construct an integral flow f' and node integers π' satisfying the same conditions with $\pi'(t) = p + 1$ (so that f' solves the $(p + 1)^{\text{st}}$ related program). Let the admissible arcs for this cycle be those for which

$$(3.24) \qquad \pi(y) - \pi(x) = a(x, y).$$

(Thus if $a(x, y) > 0$, at most one of (x, y) and (y, x) is admissible. For the starting point $\pi = 0$, only arcs with $a(x, y) = 0$ are admissible.)

Minimal cost flow routine. Perform the labeling process on the subnetwork of admissible arcs, beginning with the flow f. Thus the labeling rules are: node y can be labeled from node x if either

(a) (x, y) is admissible and $f(x, y) < c(x, y)$,

(b) (y, x) is admissible and $f(y, x) > 0$.

If breakthrough occurs, change the flow in the usual way, and re-label. If non-breakthrough occurs, denote the present flow (which may or may not be f) by f' and define new node integers by

$$(3.25) \qquad \pi'(x) = \begin{cases} \pi(x), & x \in X, \\ \pi(x) + 1, & x \in \overline{X}, \end{cases}$$

where X is the set of labeled nodes.

The routine is then repeated, using the new node integers π' to define admissible arcs, and beginning with the flow f'. Eventually a maximal flow is constructed, as we shall show later. First, we point out that if the flow f and node numbers π satisfy (3.18)–(3.20), then the same node numbers and flow f' do also, simply because the routine does not permit flow changes in inadmissible arcs. In addition, f' and π' satisfy these conditions with $\pi'(t) = p + 1$, that is

$$(3.26) \qquad \pi'(s) = 0, \pi'(t) = p + 1,$$

116

(3.27) $\qquad \pi'(y) - \pi'(x) < a(x, y) \Rightarrow f'(x, y) = 0,$
(3.28) $\qquad \pi'(y) - \pi'(x) > a(x, y) \Rightarrow f'(x, y) = c(x, y).$

Condition (3.26) is obvious from (3.25) and the fact that $s \in X$, $t \in \bar{X}$ if non-breakthrough occurs. To establish (3.27), suppose $\pi'(y) - \pi'(x) < a(x, y)$. Then from (3.25) and the fact that we are dealing with integers, we have $\pi(y) - \pi(x) \leqslant a(x, y)$. If strict inequality holds, then (x, y) was inadmissible throughout the routine, and hence $f'(x, y) = f(x, y) = 0$. If, on the other hand, equality holds, then from (3.25) we must have $\pi'(y) = \pi(y)$, $\pi'(x) = \pi(x) + 1$, and hence $x \in \bar{X}$, $y \in X$ at the conclusion of the labeling process. Since (x, y) was admissible, it follows that $f'(x, y) = 0$, as otherwise x would be labeled from y (labeling rule (b)), a contradiction. Thus $f'(x, y) = 0$ in either case, establishing (3.27).

The proof of (3.28) is similar.

Consequently, if we denote the outputs of the pth application of the routine by f_p and π_p, we may state the following theorem and corollaries.

THEOREM 3.1. *The node integers π_p and integral flow f_p satisfy the optimality properties (3.18), (3.19), (3.20) for the pth related program. In addition, so do π_p and f_{p+1}.*

COROLLARY 3.2. *The flows f_p and f_{p+1} maximize the linear form $pv - \sum_{\mathcal{A}} a(x, y)f(x, y)$ (where v is the value of f) over all flows from source to sink. Moreover, the corresponding node integers π_p and arc integers $\gamma_p(x, y)$ $= \max(0, \pi_p(y) - \pi_p(x) - a(x, y))$ solve the dual program. In particular,*

$$pv_p - \sum_{\mathcal{A}} a(x, y)f_p(x, y) = pv_{p+1} - \sum_{\mathcal{A}} a(x, y)f_{p+1}(x, y)$$

(3.29)
$$= \sum_{\mathcal{A}} c(x, y)\gamma_p(x, y).$$

It follows from Corollary 3.2 that f_p minimizes total flow cost over all flows of value v_p. It is also clear from this corollary that for all sufficiently large p, f_p must be a maximal flow. One way of making this last statement more precise is by introducing the notion of "path cost." Consider any path from s to t (I.1) and sum the arc costs for all forward arcs of the path, then subtract from this the sum of arc costs corresponding to reverse arcs of the path. We call the resulting number the *path cost*.

COROLLARY 3.3. *If p is greater than the maximal path cost from s to t, then f_p is maximal and minimizes cost over all maximal flows.*

PROOF. If f_p is not maximal, then there is a flow augmenting path from s to t (Corollary I.5.2); that is, there is a flow f of value $v = v_p + \varepsilon$ ($\varepsilon > 0$) and a path from s to t such that

$$f(x, y) = \begin{cases} f_p(x, y) + \varepsilon, & \text{if } (x, y) \text{ is a forward arc of the path,} \\ f_p(x, y) - \varepsilon, & \text{if } (x, y) \text{ is a reverse arc of the path,} \\ f_p(x, y), & \text{otherwise.} \end{cases}$$

117

Hence, letting p' be the path cost, we have

$$\sum_{\mathscr{A}} a(x, y)[f(x, y) - f_p(x, y)] = p'\varepsilon < p\varepsilon = p(v - v_p)$$

or

$$pv_p - \sum_{\mathscr{A}} a(x, y)f_p(x, y) < pv - \sum_{\mathscr{A}} a(x, y)f(x, y),$$

contradicting Corollary 3.2. Thus f_p is maximal, and hence minimizes cost over all maximal flows.

Once a maximal flow has been constructed by the routine, the original problem has either been solved or shown to be infeasible. Thus Corollary 3.3 gives a bound on the computation. In terms of the number of individual labelings, this bound would involve the total demand and the cost function. As was the case for the Hitchcock problem, a bound can be obtained that depends only on the total demand and the number of nodes. The idea is the same: one looks at consecutive occurrences of non-breakthroughs between breakthroughs. If we put the algorithm in slightly different terms, the difference being that trivial non-breakthrough situations are thrown out by making a large enough node number change to ensure that at least one more node will be labeled on the next labeling, then such a bound is obtained easily.

To see what the node number change should be to guarantee this situation, we need only examine arcs of the sets $(X, \bar{X})$ and $(\bar{X}, X)$, since adding a constant to $\pi(x)$ for $x \in \bar{X}$ does not change the pattern of admissibility of arcs of the sets (X, X) and $(\bar{X}, \bar{X})$. Thus we consider the following six cases, the first three of which correspond to arcs of $(X, \bar{X})$, the last three, to arcs of $(\bar{X}, X)$:

(a) $\pi(\bar{x}) - \pi(x) = a(x, \bar{x})$ (hence $f(x, \bar{x}) = c(x, \bar{x})$),

(b) $\pi(\bar{x}) - \pi(x) > a(x, \bar{x})$ (hence $f(x, \bar{x}) = c(x, \bar{x})$),

(c) $\pi(\bar{x}) - \pi(x) < a(x, \bar{x})$ (hence $f(x, \bar{x}) = 0$),

(d) $\pi(x) - \pi(\bar{x}) = a(\bar{x}, x)$ (hence $f(\bar{x}, x) = 0$),

(e) $\pi(x) - \pi(\bar{x}) > a(\bar{x}, x)$ (hence $f(\bar{x}, x) = c(\bar{x}, x)$),

(f) $\pi(x) - \pi(\bar{x}) < a(\bar{x}, x)$ (hence $f(\bar{x}, x) = 0$).

If we add a small constant δ to $\pi(\bar{x})$, e.g., $\delta = 1$, then possible changes in the admissibility structure are indicated by the diagram:

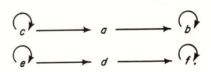

It follows that if we define $\bar{a}(x, y)$ by (3.22), we may determine a node number change δ as follows. First single out the subsets of arcs corresponding to (c) and (e) above:

$$(3.30) \qquad \mathscr{A}_1 = \{(x, y) | x \in X, y \in \overline{X}, \bar{a}(x, y) > 0\},$$

$$(3.31) \qquad \mathscr{A}_2 = \{(x, y) | x \in \overline{X}, y \in X, \bar{a}(x, y) < 0\}.$$

Let

$$(3.32) \qquad \delta_1 = \min_{\mathscr{A}_1} [\bar{a}(x, y)],$$

$$(3.33) \qquad \delta_2 = \min_{\mathscr{A}_2} [-\bar{a}(x, y)].$$

Then the node number change

$$(3.34) \qquad \delta = \min (\delta_1, \delta_2) > 0$$

introduces at least one more admissible arc from one of the sets $\mathscr{A}_1$ or $\mathscr{A}_2$. (Moreover, the optimality properties (3.18), (3.19), (3.20) again hold for the old flow f and new node integers.) Consequently the old labeling can be repeated, and in addition at least one more node can be labeled. Since the source s is always labeled, the maximal number of consecutive non-breakthroughs cannot exceed the number of nodes in the network.

Admissible arcs corresponding to (a) and (d) above become inadmissible for the next labeling.

We note the following corollary.

COROLLARY 3.4. *The flow $f_{p+\delta}$ maximizes the linear form $p'v - \sum a(x, y) f(x, y)$ for all p' in the interval $p \leqslant p' \leqslant p + \delta$. Here δ is the node number change (3.34), and $f_{p+\delta}$ is the flow that produces non-breakthrough and the subsequent change δ in the minimal cost flow routine.*

Termination of the computation is recognized when the minimizing sets in (3.32) and (3.33) are both empty, which is equivalent to saying that every arc of $(X, \overline{X})$ is saturated, whereas every arc of $(\overline{X}, X)$ is flowless. Thus $(X, \overline{X})$ is a minimal cut and the flow is maximal.

The amount by which the dual form $\sum c(x, y)\gamma(x, y)$ changes with an occurrence of non-breakthrough can be determined directly, or can be found from Corollary 3.4 as follows. Since $f_{p+\delta}$ maximizes $pv - \sum a(x, y) f(x, y)$, we have

$$\sum c(x, y)\gamma_p(x, y) = pv_{p+\delta} - \sum a(x, y)f_{p+\delta}(x, y).$$

Since $f_{p+\delta}$ also maximizes $(p + \delta)v - \sum a(x, y) f(x, y)$, we have

$$(p + \delta)v_{p+\delta} - \sum a(x, y) f_{p+\delta}(x, y) = \sum c(x, y)\gamma_{p+\delta}(x, y).$$

Hence

$$\sum c(x, y)\gamma_p(x, y) = \sum c(x, y)\gamma_{p+\delta}(x, y) - \delta v_{p+\delta}.$$

Thus the dual form increases by $\delta v_{p+\delta}$ in going from related problem p to $p + \delta$.

The discussion thus far has centered around the maximization of the linear form (3.6), or rather the family of forms (3.6) for variable p. But there is another, and more basic, way of viewing the method of this section that should be emphasized. We noted following Corollary 3.2 that the flow f_p minimizes cost over all flows from source to sink that have value $v_p = v(f_p)$. Suppose we let a_p be the cost of f_p,

$$a_p = \sum_{\mathscr{A}} a(x, y) f_p(x, y).$$

We have

$$0 = v_0 \leqslant v_1 \leqslant v_2 \leqslant \ldots \leqslant v_P,$$
$$0 = a_0 \leqslant a_1 \leqslant a_2 \leqslant \ldots \leqslant a_P,$$

and thus the sequence of points (v_p, a_p), $p = 0, 1, \ldots, P$, generates a piecewise linear, monotonic increasing curve $a(v)$ in the (v, a)-plane by joining distinct, adjacent points of the sequence with line segments. It follows that $a(v)$ is the minimal flow cost corresponding to the flow value v for all v satisfying $0 \leqslant v \leqslant v_P$. For suppose not. Then there would be a v satisfying $v_p < v < v_{p+1}$, for some $p = 0, 1, \ldots, P - 1$, and a flow $\tilde{f}$ from source to sink of value v with cost $\tilde{a}(v) < a(v)$. Let $v = \alpha v_p + (1 - \alpha)v_{p+1}$, $0 < \alpha < 1$, and consider the flow $f = \alpha f_p + (1 - \alpha)f_{p+1}$ of value v and cost $a(v)$. By assumption,

$$pv - \tilde{a}(v) > pv - a(v) = pv - \sum a(x, y)f(x, y),$$

and hence upon substituting for v and f,

$$pv - \tilde{a}(v) > \alpha[pv_p - a(v_p)] + (1 - \alpha)[pv_{p+1} - a(v_{p+1})].$$

By Corollary 3.2, the bracketed terms here are equal, and so

$$pv - \tilde{a}(v) > pv_p - a(v_p),$$

contradicting the fact that f_p maximizes the form $pv - \sum a(x, y)f(x, y)$ over all flows from source to sink. Thus the function $a(v)$ generated in the computation gives the minimal cost for $0 \leqslant v \leqslant v_P$. A consequence of this fact is that $a(v)$ is convex.

Notice that this viewpoint provides a simple conceptual framework for the computational process, which might be summarized by the statement: each additional unit of flow through the network travels by a least cost flow augmenting path. In more detail, suppose we have constructed a flow f of value v at some stage of the algorithm. The procedure then finds a flow augmenting path with respect to f that has the least path cost of all such paths. Thus each linear piece of $a(v)$ corresponds to an interval in which the flow augmenting paths that are located have a constant path cost, given by the slope of $a(v)$. If $v_p < v_{p+1}$, the slope of $a(v)$ in the interval (v_p, v_{p+1}) is equal to p, since by (3.29),

$$a(v_{p+1}) - a(v_p) = p(v_{p+1} - v_p).$$

(This can also be seen directly by summing the equations $\pm [a(x, y) + \pi_p(x) - \pi_p(y)] = 0$ along a path from s to t consisting of admissible arcs, the plus sign being taken for forward arcs, the minus sign for reverse arcs, to obtain the result that the path cost is p.)

The foregoing discussion suggests the validity of Theorem 3.5 below, which has been stated by Jewell [48] and Busacker and Gowen [3]; it is also implicit in the computational procedure described by Iri [45]. This theorem may properly be regarded as the central one concerning minimal cost flows.

THEOREM 3.5. *Let f be a minimal cost flow from source to sink corresponding to the flow value v. Then the flow f' obtained from f by adding $\varepsilon > 0$ to the flow in forward arcs of a minimal cost f-augmenting path, and subtracting ε from the flow in reverse arcs of the path, is a minimal cost flow corresponding to the value $v + \varepsilon$.*

Note that Theorem 3.5 strengthens the principle brought out above by allowing f to be any minimal cost flow of value v. The proof given here simply uses the fact that such an f and some π_p must satisfy the optimality properties (3.18), (3.19), (3.20).

PROOF. We may assume that v satisfies $v_p \leqslant v < v_{p+1}$. Then every f-augmenting path has cost $\geqslant p$. To prove the theorem it suffices to show that equality holds here for some f-augmenting path.

We first note that f and π_p satisfy the optimality properties (3.19), (3.20) (and of course (3.18)) for the pth related program. For, letting $\bar{a}(x, y) = a(x, y) + \pi_p(x) - \pi_p(y)$, we have from (3.23),

$$\sum \bar{a}(x, y)f_p(x, y) = a(v_p) - pv_p,$$

$$\sum \bar{a}(x, y)f(x, y) = a(v) - pv.$$

Now (3.19), (3.20) hold for f_p and π_p; if they did not also hold for f and π_p, then clearly

$$\sum \bar{a}(x, y)f(x, y) > \sum \bar{a}(x, y)f_p(x, y),$$

and hence from the last two displayed equations,

$$a(v) - a(v_p) > p(v - v_p),$$

a contradiction.

Assume now that every f-augmenting path has path cost $> p$. Consider the labeling process of the algorithm applied to f and the sub-network of

121

π_p-admissible arcs. Non-breakthrough must result, since all paths of admissible arcs from s to t have cost p. It follows that the node numbers π'_{p+1} obtained from π_p by adding unity to $\pi_p(x)$ for unlabeled x, together with the flow f, satisfy the optimality properties for the $(p+1)^{\text{st}}$ related program. Thus

$$pv - a(v) = pv_p - a(v_p) = pv_{p+1} - a(v_{p+1})$$

$$(p+1)v - a(v) = (p+1)v_{p+1} - a(v_{p+1}),$$

whence subtracting gives $v = v_{p+1}$, contradicting the assumption $v < v_{p+1}$. Thus some f-augmenting path has cost p, as was to be shown.

It follows from Theorem 3.5 that any method for finding a minimal cost flow augmenting path can be used to solve the minimal cost flow problem posed in this section, since one can always start with the zero flow. The method we have described is merely one such. (For others, see [3] and [45], for example.) Conceptually it is not the simplest, but it has certain advantages for later use.

It should also be remarked that the validity of Theorem 3.5 does not actually depend on the assumption of non-negative costs that has been made in this section. The reason for assuming non-negative costs here is that it is then easy to initiate the computational routine, e.g., with $f = 0$, $\pi = 0$. Later we will describe (§ 11) a more general and flexible method requiring no assumption concerning the cost function.

We conclude the discussion of this section with the following theorem, which will be invoked in the study of maximal dynamic flows in § 9.

THEOREM 3.6. *Let $x_1, x_2, \ldots, x_k$ ($x_1 = s$, $x_k = t$) be any chain from s to t. Then*

$$(3.35) \qquad \sum_{i=1}^{k-1} [a(x_i, x_{i+1}) + \gamma_p(x_i, x_{i+1})] \geqslant p.$$

If $f_p(x_i, x_{i+1}) > 0$ for all arcs of the chain, then $\gamma_p(x_i, x_{i+1}) > 0$ for some arc of the chain, and equality holds in (3.35).

PROOF. The first assertion results from summing the inequalities

$$\pi_p(x_i) - \pi_p(x_{i+1}) + \gamma_p(x_i, x_{i+1}) \geqslant -a(x_i, x_{i+1})$$

along the chain, and noting that $\pi_p(s) = 0$, $\pi_p(t) = p$.

If $f_p(x_i, x_{i+1}) > 0$, then by Theorem 3.1, equality holds in the last displayed inequality, hence also in (3.35) if every arc of the chain has positive flow.

For the remaining assertion of the theorem, assume that $f_p(x_i, x_{i+1}) > 0$ ($i = 1, \ldots, k-1$) and consider the labeling process using f_p that resulted in non-breakthrough. Since $s \in X$ and $t \in \overline{X}$, there is a node x_m of the chain with $x_m \in X$, $x_{m+1} \in \overline{X}$. If $f_{p-1}(x_m, x_{m+1}) > 0$, then by Theorem 3.1,

$$\pi_{p-1}(x_{m+1}) - \pi_{p-1}(x_m) \geqslant a(x_m, x_{m+1}).$$

If, on the other hand, $f_{p-1}(x_m, x_{m+1}) = 0$, then, since the flow in arc (x_m, x_{m+1}) has changed, the arc was admissible for the labeling process; that is,

$$\pi_{p-1}(x_{m+1}) - \pi_{p-1}(x_m) = a(x_m, x_{m+1}).$$

Now

$$\pi_p(x_m) = \pi_{p-1}(x_m),$$

$$\pi_p(x_{m+1}) = \pi_{p-1}(x_{m+1}) + 1.$$

Hence, since we have established

$$\pi_{p-1}(x_{m+1}) - \pi_{p-1}(x_m) - a(x_m, x_{m+1}) \geqslant 0,$$

it follows that

$$\gamma_p(x_m, x_{m+1}) = \pi_p(x_{m+1}) - \pi_p(x_m) - a(x_m, x_{m+1}) > 0.$$

EXAMPLE. To illustrate the construction of a maximal flow that minimizes cost, consider the network of Fig. 3.1, in which the first number

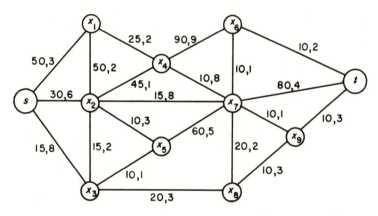

Figure 3.1

on an arc is its capacity; the second, its unit shipping cost. We assume that the problem is an undirected one, the given costs and capacities holding for both directions.

If we begin with all node numbers zero and proceed with the computation, no positive flow gets through the network until $\pi(t) = 15$. The node integers π_{15} are shown in Fig. 3.2. From these, admissible arcs are determined (indicated by black arrowheads in Fig. 3.2), and the labeling

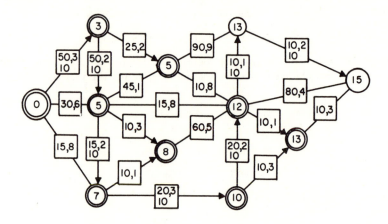

Figure 3.2 (π_{15}, f_{16})

process (using $f_{15} = 0$) yields the flow f_{16} (shown in the lower left-hand corners of the boxes in Fig. 3.2) together with the final labeled set X for the cycle (the double circles of Fig. 3.2). Here $\delta = 1$, new admissible arcs are (x_4, x_6), (x_7, t), (x_9, t), whereas (x_7, x_6) becomes inadmissible by virtue of having a positive arc number $\gamma_{16}(x_7, x_6) = 1$ (shown in the lower right-hand corner of the box in Fig. 3.3; white arrowheads are used to indicate

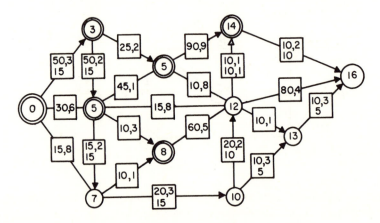

Figure 3.3 (π_{16}, f_{17})

flow directions in such arcs). Starting with f_{16} of Fig. 3.2, and labeling on the admissible sub-network of Fig. 3.3, produces the flow f_{17} shown in Fig. 3.3, and finally the labeled set X for the cycle.

124

Figures 3.4–3.9 show the subsequent outputs of the routine. The flow f_{23} of Fig. 3.9 is maximal (a minimal cut being shown by the heavy arcs) and of course minimizes cost over all maximal flows.

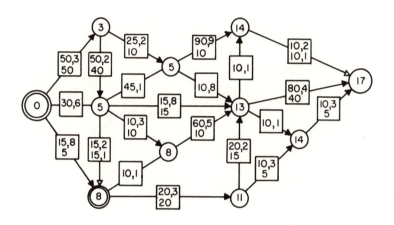

Figure 3.4 (π_{17}, f_{18})

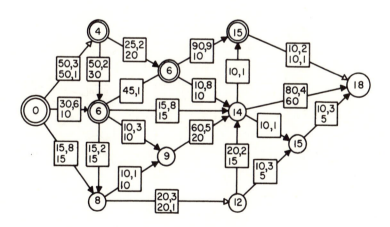

Figure 3.5 (π_{18}, f_{19})

125

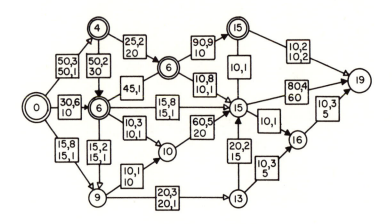

Figure 3.6 (π_{19}, f_{20})

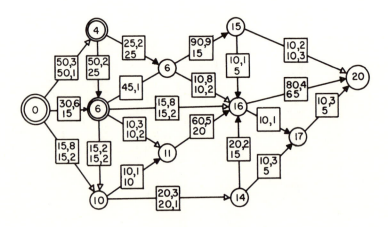

Figure 3.7 (π_{20}, f_{21})

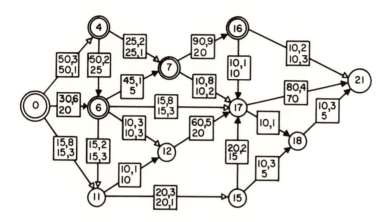

Figure 3.8 (π_{21}, f_{22})

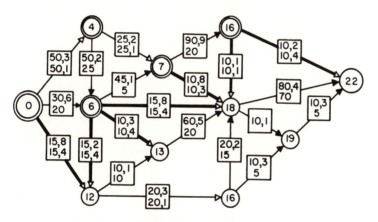

Figure 3.9 (π_{22}, f_{23})

Observe the behavior of the (undirected) arc (x_7, x_6) throughout the computation. Initially, for small values of p, it carries flow from x_7 to x_6; eventually the arc becomes admissible in the other direction, and is saturated in this direction in the final flow.

4. Equivalence of Hitchcock and minimal cost flow problems

It is clear that the Hitchcock problem is a special case of the minimal cost flow problem. It is not so obvious that the reverse is true. That the two problems are equivalent, however, can be established in various ways. For instance, a device due to Orden [62] can be used to pass from a capacity constrained trans-shipment problem to a capacity constrained Hitchcock problem; one can then apply a technique due to Dantzig [8] to convert the

127

latter to a standard Hitchcock problem. We shall not, in fact, follow this path. Rather we shall exploit a method suggested by Wagner [70] and make the transition in one step.

We first remark that it entails no loss of generality to take the general flow problem in equation form, that is

(4.1) $$f(x, N) - f(N, x) = a(x), \qquad x \in S,$$

(4.2) $$f(x, N) - f(N, x) = 0, \qquad x \in R,$$

(4.3) $$f(N, x) - f(x, N) = b(x), \qquad x \in T,$$

(4.4) $$0 \leqslant f(x, y) \leqslant c(x, y), \qquad (x, y) \in \mathscr{A},$$

(4.5) $$\text{minimize} \sum_{\mathscr{A}} a(x, y) f(x, y).$$

For if (4.1) and (4.3) are inequalities (respectively $\leqslant$ and $\geqslant$), we may insert an additional sink t together with the arcs (S, t), each of these having large capacity and zero cost, and place a demand at the sink equal to $a(S) - b(T)$, thereby obtaining an equivalent problem in equation form.†

To convert the problem (4.1)–(4.5) to a Hitchcock problem, define a bipartite network from the given one $[N; \mathscr{A}]$ as follows. Each source in the new network corresponds to an arc of the old; we denote them by ordered pairs x, y; each sink in the new network corresponds to a node of the old. All arcs of the new network lead from sources to sinks: source x, y is connected to sink x with cost $a(x, y; x) = 0$, and to sink y with cost $a(x, y; y) = a(x, y)$. The "other arcs" are missing, or may be assumed to have infinite cost. (Strictly speaking, we should take the latter point of view, since we have defined the Hitchcock problem as if all arcs from sources to sinks are present. It is merely a convenience of exposition, however, to throw out arcs unless they have either the form $(x, y; x)$ or $(x, y; y)$, and we shall do this.) At source x, y place a supply $c(x, y)$; at sink x put a demand $c(x, N) - a(x)$, $c(x, N)$, or $c(x, N) + b(x)$ according as $x \in S$, $x \in R$, or $x \in T$ in the old network. Note that $c(x, N) - a(x) \geqslant 0$ if the original problem is feasible, and that balance of supply and demand in the original problem implies the same for the new problem.

EXAMPLE. If the original problem is that schematized in Fig. 4.1, then the new problem is pictured in Fig. 4.2.

---

† If negative costs are present, this specific translation to a problem in equation form may not be valid. However, if one also adds the arcs (T, t), each with large capacity and zero cost, an equivalent flow problem is always obtained. Then the Hitchcock problem into which this is converted by the method of this section will also be in equation form, but will have negative costs if the original problem does. If desired, these costs a_{ij} can be made non-negative by changing them to $a_{ij} - \min_i a_{ij}$, since for a Hitchcock problem in equation form, any cost transformation $\bar{a}_{ij} = a_{ij} + \alpha_i - \beta_j$ yields an equivalent problem.

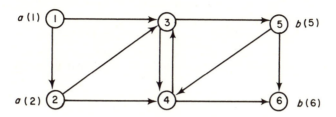

Figure 4.1

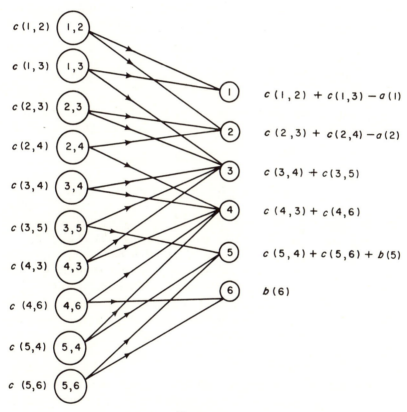

Figure 4.2

The new problem is to determine arc flows $f(x, y; z)$ (thus $f(x, y; z)$ is defined for $(x, y) \in \mathcal{A}$, and $z = x$ or $z = y$), that solve the program

(4.6) $$f(x, y; y) + f(x, y; x) = c(x, y),$$

(4.7) $$\sum_{y \in N} [f(x, y; x) + f(y, x; x)] = c(x, N) + \begin{cases} -a(x), & x \in S, \\ 0, & x \in R, \\ b(x), & x \in T, \end{cases}$$

129

$$(4.8) \qquad\qquad f(x, y; z) \geqslant 0,$$

$$(4.9) \qquad\qquad \text{minimize} \sum_{(x,y) \in \mathscr{A}} a(x, y; y) f(x, y; y).$$

Suppose now that $f(x, y)$ is feasible for the original problem. Then define

$$(4.10) \qquad\qquad f(x, y; y) = f(x, y),$$

$$(4.11) \qquad\qquad f(x, y; x) = c(x, y) - f(x, y).$$

Thus arc flows in the new problem are non-negative. In addition, (4.6) and (4.7) are satisfied, since

$$f(x, y; y) + f(x, y; x) = c(x, y),$$

$$\sum_{y} [f(x, y; x) + f(y, x; x)] = c(x, N) - f(x, N) + f(N, x).$$

Conversely, if the new problem is feasible, and if we define $f(x, y)$ by (4.10), it is clear that the constraints (4.4) are satisfied. Also,

$$f(x, N) - f(N, x) = \sum_{y} [f(x, y; y) - f(y, x; x)]$$

$$= \sum_{y} [c(x, y) - f(x, y; x)] - \sum_{y} f(y, x; x)$$

by (4.6). Using (4.7), the right side of this equality reduces to $a(x)$, 0, or $-b(x)$ according as $x \in S$, $x \in R$, or $x \in T$, verifying (4.1), (4.2), (4.3).

Since it is also clear that corresponding feasible flows in the two problems have the same cost, it follows that a general minimal cost flow problem can be transformed to a Hitchcock problem in this way.

Thus it suffices to confine attention to bipartite graphs in studying flow problems. It would be pointless to do so, however, both from the mathematical and computational viewpoints.

5. A shortest chain algorithm

A special minimal cost flow problem having independent interest is that of finding a minimal cost (or shortest) chain from one node to another in a network in which each arc (x, y) has an associated cost (or length) $a(x, y)$ $\geqslant 0$. While this is a purely combinatorial problem, it may also be considered as a minimal cost flow problem by placing unit supply at the first node (the source) and unit demand at the second node (the sink), taking arc capacities infinite, and asking for a feasible flow that minimizes cost. Since the algorithm of § 3 constructs an integral flow, it solves the shortest chain problem. In other words, the first unit of flow constructed by the algorithm travels by a least cost chain.

§5. A SHORTEST CHAIN ALGORITHM

Other ways of solving the shortest chain problem have been proposed [1, 9, 20, 58]. We mention one that has been suggested by Minty for the case of undirected networks [58]. Simply build a string model of the network, the lengths of the pieces of string being proportional to the given arc lengths, take the source in one hand, the sink in the other, and stretch. Thus one is solving the given minimum problem by maximizing. Indeed, the analogue device solves the dual maximum problem.

Applications of the shortest chain problem are numerous. Some come to mind readily. For instance, in making up a table of highway distances between cities, a shortest chain between each pair of cities needs to be computed. Or, in setting up a Hitchcock problem, it is frequently the case that many alternate shipping routes exist between a given source and sink. Then either a minimal cost route needs to be computed, or the problem should be formulated directly as one of trans-shipment. Another problem that has been viewed, in its discrete form, as a shortest chain problem, is that of determining the least time for an airplane to climb to a given altitude [5].

Our purpose in this section is to describe a combinatorial method for the shortest chain problem that works under a less restrictive assumption than $a(x, y) \geqslant 0$, the less restrictive assumption being: the sum of costs around any directed cycle is non-negative.† (For an undirected network, this assumption is no less restrictive: it implies $a \geqslant 0$.)

As in other methods for solving the shortest chain problem, the method described below does more: it yields, on one application, shortest chains from the source to all other nodes of the network that can be reached from the source by chains. Thus, for example, if one were faced with the problem, mentioned previously, of finding shortest highway distances between each pair of some set of cities, it would not be necessary to repeat the algorithm for each pair.

In essence, the algorithm works with the dual of the shortest chain problem, although it is not necessary, in the proof, to use this fact explicitly.

Shortest chain algorithm.

(1) Start by assigning all nodes labels of the form $[-, \pi(x)]$, where $\pi(s) = 0$, $\pi(x) = \infty$ for $x \neq s$.

(2) Search for an arc (x, y) such that $\pi(x) + a(x, y) < \pi(y)$. (Here $\infty + a = \infty$.) If such an arc is found, change the label on node y to $[x, \pi(x) + a(x, y)]$, and repeat. (That is, the new $\pi(y)$ is $\pi(x) + a(x, y)$.) If no such arc is found, terminate.

---

† It can be shown that if no assumption is made about the function $a(x, y)$, then the shortest chain problem is equivalent to the "traveling salesman problem" [12, 40], for which no really simple algorithms are known. Thus some such assumption as non-negative directed cycle costs appears essential here.

131

At any stage of the computation, if $\pi(y) < \infty$ for $y \neq s$, then y has a label whose first member is some node x, and $\pi(x) + a(x, y) \leq \pi(y)$, hence $\pi(x) < \infty$. For the existence of the label $[x, \pi(y)]$ implies that at some (possibly earlier) stage, $\pi(x) + a(x, y) = \pi(y)$; and while $\pi(x)$ may be later reduced to produce an inequality, if $\pi(y)$ were also reduced later, either y is not labeled from x, or else equality holds again. It follows that if, at any stage, one starts at such a node y and follows the labels, then one eventually either

(a) arrives at s and must stop, or

(b) cycles.

Suppose that at every stage of the calculation, case (a) above holds for every y with $\pi(y) < \infty$. Let

$$s = x_1, x_2, \ldots, x_n = y$$

be the chain from s to y singled out by the labels at any stage. Then

(5.1) $$\pi(x_i) + a(x_i, x_{i+1}) \leq \pi(x_{i+1}), \qquad i = 1, \ldots, n - 1,$$

and summing these inequalities along the chain gives

$$\pi(s) + \sum_{i=1}^{n-1} a(x_i, x_{i+1}) \leq \pi(y).$$

But $\pi(s) = 0$, for otherwise s would be labeled from some node, and case (a) would not hold. Thus

(5.2) $$\pi(y) \geq \sum_{i=1}^{n-1} a(x_i, x_{i+1}).$$

It follows that the computation terminates, since chain lengths are bounded below, and the node numbers $\pi(y)$ are monotone decreasing. At termination, (5.1) must be an equality, and hence so is (5.2). Suppose that, at termination, there were a shorter chain from s to y than the one yielded by the labels, say

$$s = x_1', x_2', \ldots, x_m' = y.$$

Then, for this chain

$$\pi(x_1') + \sum_{i=1}^{m-1} a(x_i', x_{i+1}') < \pi(x_m').$$

But, since termination has occurred, we have

$$\pi(x_i') + a(x_i', x_{i+1}') \geq \pi(x_{i+1}'), \qquad i = 1, \ldots, m - 1,$$

whence summing gives a contradiction.

Now suppose that at some stage of the computation, case (b) above holds. Let

$$x_1, x_2, \ldots, x_k = x_1$$

be a directed cycle yielded by the labels, and suppose that x_j was the last node of the cycle to receive a label from its predecessor. At this stage (immediately before x_j was labeled from x_{j-1}) we had

$$\pi(x_j) > \pi(x_{j-1}) + a(x_{j-1}, x_j),$$
$$\pi(x_{j+1}) \geqslant \pi(x_j) + a(x_j, x_{j+1}).$$

Letting primes denote the node numbers after labeling x_j from x_{j-1}, it follows that

$$\pi'(x_{j+1}) > \pi'(x_j) + a(x_j, x_{j+1}).$$

Hence, summing the inequalities

$$\pi'(x_{i+1}) \geqslant \pi'(x_i) + a(x_i, x_{i+1})$$

around the directed cycle, and noting that one of them is strict, we have

$$\sum_{i=1}^{k-1} a(x_i, x_{i+1}) < 0.$$

Thus, under our assumption that no directed cycles have negative lengths, case (b) cannot occur, the algorithm terminates, and at termination, shortest chains have been located from s to all other nodes of the network that can be reached from s by chains. Moreover, the node number for such a node is the length of a shortest chain.

The shortest chain algorithm can be used to obtain a better starting dual solution for the minimal cost flow problem of § 3. That is, instead of starting with all node numbers zero, one can find a shortest chain from source to sink by the method of this section, and then use the resulting node numbers, together with the zero flow, to initiate the computation of a maximal flow that minimizes cost.

The algorithm can also be used if costs are allowed to be negative, so long as directed cycle costs are non-negative, to find node numbers which, together with the zero flow, can be used to start the minimal cost routine. The computation from there on is no different, so long as it is desired to find a maximal flow that minimizes cost, and the theory of § 3 is unchanged. Not only is this so, but the shortest chain method can also be used to find a minimal cost flow augmenting path with respect to a given minimal cost flow of value v, and hence, in view of Theorem 3.5, the problem of § 3 can be solved by repeated applications of this method.

To see this, suppose we have constructed a minimal cost f of value v and wish to find a minimal cost f-augmenting path. One way of doing this is to

133

define a new network $[N; \mathscr{A}']$ from the given one $[N; \mathscr{A}]$, and an appropriate length function $a'(x, y)$ for (x, y) in $\mathscr{A}'$, as follows. First note that we may assume that not both $f(x, y)$ and $f(y, x)$ are positive, by virtue of $a(x, y) + a(y, x) \geqslant 0$. Now put (x, y) in $\mathscr{A}'$ if either $f(x, y) < c(x, y)$ or $f(y, x) > 0$, and define $a'(x, y)$ by

$$a'(x, y) = \begin{cases} a(x, y), & \text{if } f(x, y) < c(x, y) \text{ and } f(y, x) = 0 \text{ (or is undefined)}, \\ -a(y, x), & \text{if } f(y, x) > 0. \end{cases}$$

(5.3)

It follows that a chain from s to t in $[N; \mathscr{A}']$ corresponds to an f-augmenting path in $[N; \mathscr{A}]$, and vice versa. Moreover, the a'-length of the chain is equal to the a-cost of the path. It can also be seen that since f is a minimal cost flow, the function a' satisfies the non-negative directed cycle condition. Hence the routine of this section can be used to construct minimal cost flows of successively higher values, since one can start with the zero flow (because of the assumption on the cost function a).

Suppose, however, that the constraints are of the form (3.1), (3.2), (3.3) and negative costs are present, but directed cycle costs are non-negative. Then it may be that an optimal solution yields strict inequalities in (3.3); that is, it may be better to oversupply certain demands, and thus imposing these demands as capacities on additional sink arcs (as was done in § 3) may not be valid. However, a problem of this kind can be dealt with computationally in various simple ways. We describe, in the next section, one such computation, a minor modification of the method of § 3.

6. The minimal cost supply-demand problem: non-negative directed cycle costs

Before describing the modifications needed in the minimal cost flow routine in order to handle a problem of form (3.1)–(3.5) with negative costs, perhaps we should first consider an example in which negative costs arise naturally. For instance, we might think of an entrepreneur faced with the following problem. In each of N successive time periods he can buy, sell, or hold for later sale some commodity, subject to the following constraints. In the ith period, there is an upper bound $a_i \geqslant 0$ on the amount of the commodity that he can purchase, an upper bound $c_i \geqslant 0$ on the amount he can hold for the next period, and a lower bound $b_i \geqslant 0$ on the amount he sells (perhaps because of prior agreements). Assuming that the entrepreneur knows buying, selling, and storage costs $\bar{p}_i \geqslant 0$, $\bar{s}_i \geqslant 0$, $\bar{h}_i \geqslant 0$, respectively, for each period, how does he maximize profit over the N periods?

One can represent this problem as that of determining a minimal cost flow from source s to sinks $t_1, \ldots, t_N$ in the network shown (for $N = 5$) in Fig. 6.1, where the demands at the sinks must be fulfilled. Thus negative

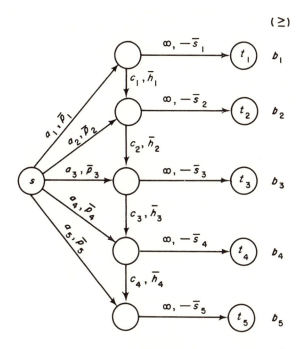

Figure 6.1

"transportation" costs can come up in a practical way. Notice too that since the representing network for this problem contains no directed cycles, the non-negative directed cycle condition is satisfied automatically.

We return to the linear program (3.1)–(3.5). The supplies $a(x)$, demands $b(x)$, and capacities $c(x, y)$ are, as usual, assumed to be positive integers.

Dual constraints for this program can be written as

$$\pi(x) - \pi(y) + \gamma(x, y) \geqslant -a(x, y),$$
$$\gamma(x, y) \geqslant 0,$$
$$\pi(x) \geqslant 0 \qquad \text{for } x \in S \cup T,$$

and consequently optimality properties for a feasible flow f and node numbers π are

(6.1) $\qquad \pi(y) - \pi(x) < a(x, y) \Rightarrow f(x, y) = 0,$

(6.2) $\qquad \pi(y) - \pi(x) > a(x, y) \Rightarrow f(x, y) = c(x, y),$

(6.3) $\quad \pi(x) \geqslant 0, x \in S; \qquad \pi(x) > 0, x \in S \Rightarrow f(x, N) - f(N, x) = a(x),$

(6.4) $\quad \pi(x) \geqslant 0, x \in T; \qquad \pi(x) > 0, x \in T \Rightarrow f(N, x) - f(x, N) = b(x).$

135

We shall show one way of using the minimal cost flow routine in order to produce f and π satisfying these properties.

As in § 3, first extend the given network $[N;\mathscr{A}]$ to $[N^*;\mathscr{A}^*]$ by adjoining nodes s, t, the arcs (s, S), (T, t), and defining

$$c(s, x) = a(x), \qquad a(s, x) = 0, \qquad\qquad x \in S,$$
$$c(x, t) = b(x), \qquad a(x, t) = 0, \qquad\qquad x \in T.$$

This extension maintains the non-negative directed cycle cost condition.

Next compute a maximal flow that minimizes cost over all maximal flows by the method of § 3, starting this time with node numbers obtained from the shortest chain algorithm, and the zero flow. Assuming that the problem is feasible, the computation terminates with a flow f for which

$$f(y, t) = c(y, t) = b(y), \qquad\qquad y \in T,$$

and node numbers $\pi(x)$ satisfying

(6.5) $$\pi(s) = 0,$$

(6.6) $$\pi(y) - \pi(x) < a(x, y) \Rightarrow f(x, y) = 0, \qquad (x, y) \in \mathscr{A}^*,$$

(6.7) $$\pi(y) - \pi(x) > a(x, y) \Rightarrow f(x, y) = c(x, y), \qquad (x, y) \in \mathscr{A}^*.$$

We now consider two cases.

Case 1. $\pi(y) \geqslant 0$ all $y \in T$. In this case the computation ends, since the restriction of f to $\mathscr{A}$ solves the original problem. To see this, we check the optimality properties (6.1)–(6.4), using π. The first two of these are obvious, since they hold in the extended network. For (6.3), we note that the computation began with node numbers of zero for $x \in S$, and thus $\pi(x) \geqslant 0$ for $x \in S$ by virtue of monotonicity. If $\pi(x) > 0$ for $x \in S$, then $\pi(x) - \pi(s) > a(s, x) = 0$, and hence $f(s, x) = c(s, x) = a(x)$ by (6.7). Thus $f(x, N) - f(N, x) = a(x)$. Property (6.4) follows from the case assumption and the termination condition $f(y, t) = b(y)$ for $y \in T$.

Case 2. $\pi(y) < 0$ for some $y \in T$. Join such y to a new sink u by arcs (y, u) with

$$c(y, u) = \infty, \qquad a(y, u) = 0.$$

Call the resulting network $[\tilde{N};\tilde{\mathscr{A}}]$. Extend f and π to $\tilde{\mathscr{A}}$ and $\tilde{N}$, respectively, by defining

$$f(y, u) = 0, \qquad \pi(u) = \min_{y \in T} \pi(y) < 0.$$

Now continue the computation, starting with π and f, and treating u as the sink, until a flow $\tilde{f}$ and associated node numbers $\tilde{\pi}$ are constructed with $\tilde{\pi}(u) = 0$. (We may think of increasing the node numbers on unlabeled nodes by unity at each non-breakthrough, stopping when $\tilde{\pi}(u) = 0$.)

Properties (6.1), (6.2), (6.3) now hold for the restrictions of $\tilde{\pi}$ and $\tilde{f}$ as in *Case* 1, and thus we need only verify (6.4). First consider those $y \in T$ for which $\pi(y) \geqslant 0$. For such y, $\tilde{\pi}(y) \geqslant 0$. Moreover, $\tilde{f}(y, t) = f(y, t) = b(y)$ for these y, and thus (6.4) holds. (Observe that the second part of the computation does not change $f(y, t)$ for any $y \in T$.) On the other hand, since the additional arcs (y, u) have infinite capacity and zero cost, it follows from (6.7) applied to $\tilde{f}$, $\tilde{\pi}$ that $\tilde{\pi}(u) - \tilde{\pi}(y) = -\tilde{\pi}(y) \leqslant 0$, and thus $\tilde{\pi}(y) \geqslant 0$ for the remaining $y \in T$, also. If $0 < \tilde{f}(y, u)$, then $\tilde{\pi}(y) = 0$ from (6.6), (6.7), and the fact that $\tilde{f}(y, u) < \infty$. Consequently, if $\tilde{\pi}(y) > 0$, then

$$0 = \tilde{f}(y, u) = \tilde{f}(N, y) - \tilde{f}(y, N) - \tilde{f}(y, t),$$

and hence

$$\tilde{f}(N, y) - \tilde{f}(y, N) = \tilde{f}(y, t) = f(y, t) = b(y).$$

7. The warehousing problem

A problem similar to, but simpler than, the entrepreneur example described in the last section is one known as the "warehousing problem" [4, 6, 10, 14, 47, 64]. Again we think of an entrepreneur who purchases, stores, and sells, in each of N successive periods, some commodity that is subject to known fluctuations in purchasing costs and selling prices. The differences between the present problem and the previous one are these:

(a) the entrepreneur has a warehouse of fixed capacity in which new purchases and hold-overs from the previous period are stored before selling;

(b) the only limitation on purchases or sales in each period is that represented by the warehouse capacity; thus supplies are infinite and demands are zero. Again profit is to be maximized.

Adopting the notation

p_i: amount purchased in period i,
w_i: amount placed in warehouse after purchasing in period i,
s_i: amount sold in period i,
h_i: amount held in warehouse after selling in period i,
c: warehouse capacity,
$\bar{p}_i$: purchase cost per unit in period i,
$\bar{w}_i$: warehouse cost per unit in period i,
$\bar{s}_i$: selling price per unit in period i,

the constraint equations and inequalities for the warehousing problem can be written as

(7.1)
$$h_{i-1} + p_i - w_i = 0, \qquad i = 1, \ldots, N; h_0 = 0,$$
$$w_i - s_i - h_i = 0, \qquad i = 1, \ldots, N; h_N = 0,$$
$$w_i \leqslant c.$$

Here all variables are non-negative. Subject to these constraints, it is desired to minimize

(7.2)
$$\sum_{i=1}^{N} (\bar{p}_i p_i + \bar{w}_i w_i - \bar{s}_i s_i).$$

(We could start with a positive initial stock h_0, and allow the possibility that the final stock h_N can be positive also. Since this makes no essential difference in the subsequent analysis, we have assumed $h_0 = h_N = 0$.)

Again we can represent this problem as a minimal cost flow problem in a suitable network; it is little different in structure from the previous one. Such a network is shown, for $N = 3$, in Fig. 7.1, with the associated arc flow variables.

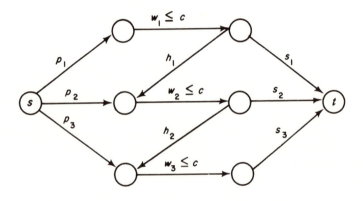

Figure 7.1

The extremely simple nature of the problem becomes more apparent through a network representation different from the one of Fig. 7.1. This representation is due to Dantzig [10]. To deduce it, we return to the constraints (7.1) that describe the problem, and replace the warehouse capacity inequality $w_i \leqslant c$ by the equation in non-negative variables

$$w_i + u_i = c.$$

Thus u_i represents the unused capacity of the warehouse in period i, and the constraints (written out for $N = 3$) appear in detached coefficient form as follows:

(7.3)

	p_1	w_1	s_1	h_1	u_1	p_2	w_2	s_2	h_2	u_2	p_3	w_3	s_3	u_3	
(1)		1			1										= c
(2)	1	-1													= 0
(3)		1	-1	-1											= 0
(4)							1			1					= c
(5)				1		1	-1								= 0
(6)							1	-1	-1						= 0
(7)												1		1	= c
(8)									1		1	-1			= 0
(9)												1	-1		= 0

If we now replace equation (7) by $-(7) + (4) - (6) - (8)$, (4) by $-(4) + (1) - (3) - (5)$, and (1) by $-(1) - (2)$ (in general, replace equation $(3n + 1)$ by $-(3n + 1) + (3n - 2) - (3n) - (3n + 2)$), and then change signs throughout, an equivalent system is obtained:

(7.4)

	p_1	w_1	s_1	h_1	u_1	p_2	w_2	s_2	h_2	u_2	p_3	w_3	s_3	u_3	
(1)	1				1										= c
(2)	-1	1													= 0
(3)		-1	1	1											= 0
(4)			-1		-1	1				1					= 0
(5)				-1		-1	1								= 0
(6)							-1	1	1						= 0
(7)								-1		-1	1			1	= 0
(8)									-1		-1	1			= 0
(9)												-1	1		= 0

The coefficient matrix of the new system has (except for the last two columns) precisely one $+1$ and one -1 in each column, the other coefficients being zero. Thus (7.4) again has a network representation, shown in Fig. 7.2, where the (redundant) equation corresponding to node 10 is the sum of all the equations of (7.4).

139

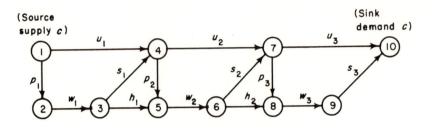

Figure 7.2

In this network representation, arcs corresponding to the variables p_i, w_i, s_i still have costs $\bar{p}_i$, $\bar{w}_i$, $-\bar{s}_i$, and we are asked to find a minimal cost flow of c units from source to sink. But since there are no capacity restrictions on arcs, and there are no directed cycles, there is an optimal flow in which all c units travel by a minimal cost chain from source to sink. Thus it suffices to find such a chain in order to solve the problem. Several facts emerge from this:

(a) The capacity of the warehouse plays no role in determining the form of an optimal solution.

(b) There is an optimal pattern of buying and selling of the all or nothing kind, that is, whatever action is taken in a period is pursued to the limit of warehouse capacity.

(c) The total profit for N periods is a multiple of the warehouse capacity.

Because of the simple structure of the representing network, an optimal policy (least cost chain) can be determined by the following trivial calculation. Start at the source, say, and compute node numbers recursively as illustrated below for $N = 3$:

$$\pi_1 = 0, \qquad \pi_2 = \pi_1 + \bar{p}_1, \qquad \pi_3 = \pi_2 + \bar{w}_1,$$
$$\pi_4 = \min(\pi_1, \pi_3 - \bar{s}_1), \qquad \pi_5 = \min(\pi_4 + \bar{p}_2, \pi_3), \qquad \pi_6 = \pi_5 + \bar{w}_2,$$
$$\pi_7 = \min(\pi_4, \pi_6 - \bar{s}_2), \qquad \pi_8 = \min(\pi_7 + \bar{p}_3, \pi_6), \qquad \pi_9 = \pi_8 + \bar{w}_3,$$
$$\pi_{10} = \min(\pi_7, \pi_9 - \bar{s}_3).$$

The π_i so determined constitute an optimal dual solution to the least cost chain problem, and by keeping a record of where the various minima occur in the calculation, a least cost chain is singled out.

8. The caterer problem

Another example of a minimal cost flow problem is the "caterer problem" [31, 46, 63]. Imagine a caterer who knows that he will require $r_j \geqslant 0$ fresh napkins on each of N successive days, $j = 1, \ldots, N$. He can meet his needs in two ways: by purchasing new napkins, or by using napkins previously laundered. However, the laundry has two kinds of service, quick and slow. A napkin sent for quick laundering is available for use m days later, whereas a napkin sent for slow service is available n days later, $0 < m < n$. New napkins purchased from the store cost $\bar{p}$ cents each, quick laundry service is $\bar{q}$ cents per napkin, and slow service $\bar{s}$ cents per napkin. How does the caterer, who starts out with no napkins, say, meet his requirements at minimal cost?

The problem perhaps appears somewhat less frivolous if stated in terms of aircraft engines and quick and slow overhaul (its actual origin), but we shall stick with napkins and laundering.

Let $p_j \geqslant 0$ represent the number of new napkins purchased for use on the jth day (remaining requirements on that day are supplied by laundered napkins), $s_j \geqslant 0$ the number sent for slow laundry service, $q_j \geqslant 0$ the number sent for quick service and $h_j \geqslant 0$ the number of soiled napkins held over to the next day. Then the problem faced by the caterer is to solve the linear program in non-negative variables:

$$(8.1) \qquad p_j + s_{j-n} + q_{j-m} \geqslant r_j, \qquad j = 1, \ldots, N,$$

$$(8.2) \qquad s_j + q_j + h_j - h_{j-1} \leqslant r_j, \qquad j = 1, \ldots, N,$$

$$(8.3) \qquad \text{minimize} \sum_{j=1}^{N} (\bar{p}_j p_j + \bar{q}_j q_j + \bar{s}_j s_j).$$

Here variables with subscripts not in the range $1, \ldots, N$ are suppressed.

The constraints (8.1) and (8.2) can be represented in network form. The representation can be made clear by considering an example with $m = 1$, $n = 2$, $N = 4$, for which a suitable network is shown in Fig. 8.1. The

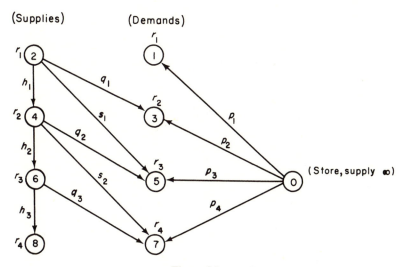

Figure 8.1

141

problem faced by the caterer is to compute a minimal cost flow that satisfies the demands at the sinks from the supplies at the sources.

9. Maximal dynamic flow

The problem taken up in this section may be stated informally as follows. Given a network $G = [N; \mathscr{A}]$ with source s and sink t, suppose that each arc of G has not only a capacity, but a traversal time as well. If, at each node of G, the commodity can either be trans-shipped immediately or held over for later shipment, determine the maximal amount of commodity flow from source to sink in a specified number of time periods [24]. For example, in the network of Fig. 9.1, the first number on an arc is its capacity in terms of commodity flow per unit time, the second number is the arc traversal time. How many units of the commodity can reach t from s in 5 time periods, say, and what is a shipping schedule that achieves this?

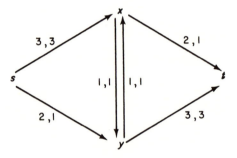

Figure 9.1

One feasible 5-period shipping schedule is shown schematically in the time-sequenced Figs. 9.2 through 9.6. Here Fig. 9.2 means that 4 units

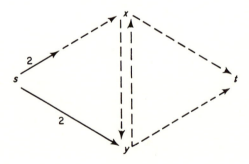

Figure 9.2

142

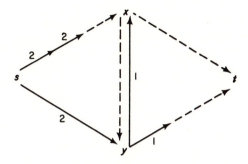

Figure 9.3

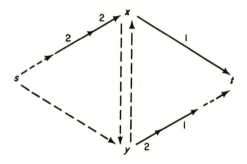

Figure 9.4

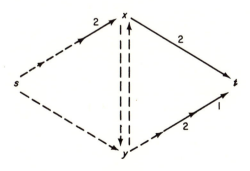

Figure 9.5

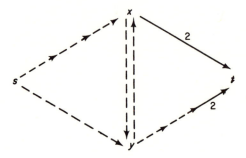

Figure 9.6

leave s at initial time 0, 2 units bound for x, the other 2 units for y; at time 1 the 2 units going to x have traversed $\frac{1}{3}$ of the arc (s, x), and the other 2 units have arrived at y. Thus Fig. 9.2 represents the time interval 0 to 1, and so on. At time 3, 1 unit arrives at t; 3 more units arrive at time 4, 4 more at time 5, giving a total flow from s to t of 8 units in the periods 0–1, 1–2, ..., 4–5. Is this a maximal dynamic flow for 5 periods, or is it possible to do better?

One can formulate the maximal dynamic flow problem as follows. Let $c(x, y)$, $a(x, y)$ be the capacity and traversal time of arc (x, y); we take these to be positive integers. Let $f(x, y; \tau)$ be the amount of flow that leaves x along (x, y) at time τ, consequently arriving at y at time $\tau + a(x, y)$. Also $f(x, x; \tau)$ is the hold-over at x from τ to $\tau + 1$. If $v(p)$ is the net flow leaving s or entering t during the p periods 0 to 1, 1 to 2, ..., $p - 1$ to p, then the problem may be stated as the linear program:

(9.1) $$\text{maximize } v(p)$$

subject to the constraints

(9.2) $$\sum_{\tau=0}^{p} \sum_{y \in N} [f(s, y; \tau) - f(y, s; \tau - a(y, s))] - v(p) = 0,$$

(9.3) $$\sum_{y \in N} [f(x, y; \tau) - f(y, x; \tau - a(y, x))] = 0,$$
$$x \neq s, t; \tau = 0, 1, \ldots, p,$$

(9.4) $$\sum_{\tau=0}^{p} \sum_{y \in N} [f(t, y; \tau) - f(y, t; \tau - a(y, t))] + v(p) = 0,$$

(9.5) $$0 \leqslant f(x, y; \tau) \leqslant c(x, y).$$

144

Here $a(x, x) = 1$, $c(x, x) = \infty$ for hold-overs at node x. It is also tacitly assumed that a variable $f(x, y; \tau)$ is suppressed if $\tau < 0$, or if, for $x \neq y$, (x, y) is not an arc of the given network G. Then the constraints (9.3) assert that for each intermediate node x and each time τ, the amount of flow that "enters" x at time τ (including the amount held over at x from time $\tau - 1$) is equal to the amount that "leaves" x at time τ (including the hold-over at x until time $\tau + 1$). Similarly (9.2) says that $v(p)$ is the net flow leaving s during the p periods, and (9.4) insists that $v(p)$ is the net flow arriving at t within the time interval. We could, of course, omit from G inward pointing arcs at s, outward pointing arcs at t.

If $f(x, y; \tau)$ and $v(p)$ satisfy (9.2)–(9.5), we call f a *dynamic flow from s to t* (for p periods) and say that f has *value v(p)*. If also $v(p)$ is maximal, then f is a *maximal dynamic flow*.

Although the constraints that describe dynamic flows may appear complicated, they are, in actuality, no more so than the simpler appearing ones for static flows. Indeed, a p-period dynamic flow through a network G corresponds to a static flow in a time-expanded version $G(p)$ of G. Here the network $G(p)$ may be constructed from G as follows. Corresponding to node x of G, $G(p)$ has $p + 1$ nodes $x(\tau)$, $\tau = 0, 1, \ldots, p$; corresponding to arc (x, y) of G, $G(p)$ has arcs $[x(\tau), y(\tau + a(x, y))]$, $0 \leqslant \tau \leqslant p - a(x, y)$; in addition, we put in arcs $[x(\tau), x(\tau + 1)]$, $0 \leqslant \tau \leqslant p - 1$, to represent hold-overs at node x. A replica $[x(\tau), y(\tau + a(x, y))]$ of (x, y) has capacity $c(x, y)$, whereas we have assumed that a hold-over arc has infinite capacity. (It will turn out that there always exists a maximal dynamic flow that avoids hold-overs at intermediate nodes, so the capacities placed on these latter are of no consequence.) If we take $s(0), s(1), \ldots, s(p)$ as sources in $G(p)$, $t(0), t(1), \ldots, t(p)$ as sinks, then the constraints characterizing a p-period dynamic flow from s to t in G are just those for a static flow from sources to sinks in $G(p)$. (In view of the existence of hold-over arcs at s and t, we could equally well take $s(0)$ as the only source, $t(p)$ as the only sink in $G(p)$.)

Figure 9.7 shows the 5-period dynamic version of the network of Fig. 9.1, together with the (static) flow in this network that corresponds to Figs. 9.2 through 9.6. All arcs are directed from left to right. Using Fig. 9.7, it can be checked that the flow shown there is maximal. A minimal cut $\mathscr{C} = (X, \overline{X})$ is given by taking the set X to consist of the nodes

$$
\begin{array}{ll}
s(\tau), & \tau = 0, 1, \ldots, 5, \\
x(\tau), & \tau = 3, 4, 5, \\
y(\tau), & \tau = 3, 4, 5.
\end{array}
$$

By expanding the network in the fashion described, the maximal dynamic flow problem can always be solved as a maximal static flow

145

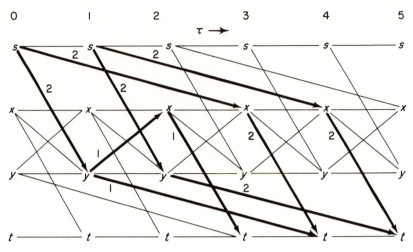

Figure 9.7

problem in the enlarged network. Moreover, it may be noted that blowing the problem up into an equivalent static problem does not require keeping arc capacities and traversal times fixed over time, as we have done. But these simplifying assumptions are essential for the much more efficient solution process to be described, which will deal only with static flows in the smaller network G.

Specifically, it will be shown that a maximal dynamic flow can always be generated from a static flow by the following device. Let f be a static flow from s to t in G that maximizes the linear function

$$(9.6) \qquad (p + 1)v - \sum_{\mathscr{A}} a(x, y)f(x, y).$$

The problem of constructing such an f has been solved in § 3. It is a simple matter to decompose f into a set of chain-flows from s to t, that is, one can easily obtain from the node arc flow f a corresponding arc-chain flow (see I.2). For example, a labeling procedure can be described to effect such a decomposition of f. Then, roughly speaking, a dynamic flow can be generated from the chain decomposition of f by starting each chain flow at time zero, and repeating each so long as there is enough time left in the p periods for the flow along the chain to arrive at the sink. This dynamic flow will be maximal for p periods.

For example, in the network G of Fig. 9.1, an f that maximizes (9.6) with $p = 4$ has a chain decomposition

$$(s, x, t; 1), \qquad (s, y, t; 1), \qquad (s, y, x, t; 1);$$

146

that is, a flow of 1 unit along each of the indicated chains. The traversal times of these chains are respectively 4, 4, 3. Consequently the first two chain flows can be repeated once in 4 periods, the last chain flow twice, giving a total flow into t of 4 units in the time interval 0 to 4. This (maximal) flow in $G(4)$ is shown in Fig. 9.8.

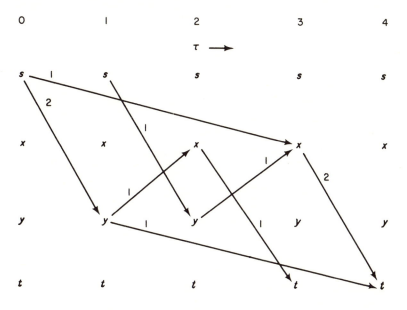

Figure 9.8

We shall call a dynamic flow (that is, a static flow in $G(p)$) that can be generated by repeating chain flows of a static flow in G, a *temporally repeated* flow. Notice that the maximal dynamic flow shown in Fig. 9.7 is not a temporally repeated flow.

The fact that there always exists a maximal dynamic flow within the subclass of temporally repeated flows is not evident. But if one knew *a priori* that this were the case, it is not difficult to see that (9.6) is the appropriate function to maximize over static flows. For if f maximizes (9.6), then (9.6) can be rewritten, in terms of a chain decomposition of f, as

$$(9.7) \qquad \sum_r (p + 1 - \sigma_r)h_r.$$

Here σ_r is the traversal time of the rth chain in the decomposition, h_r the amount of flow along this chain. (From the discussion in I.2, we would have only (9.7) $\geqslant$ (9.6). Actually equality holds if f maximizes (9.6), but in the formal proof to follow later, the inequality suffices.) Since f maximizes (9.6), it follows that $\sigma_r \leqslant p + 1$, and hence the coefficient of

147

h_r in (9.7) counts the number of times the rth chain flow can be repeated in p periods. That is, (9.7) or (9.6) is the value $v(p)$ of the temporally repeated flow generated by f.

This provides the heuristic background for examining static flows that maximize (9.6).

Suppose that f_{p+1} (notation as in § 3) has been constructed using the minimal cost flow routine discussed in § 3. Thus f_{p+1} maximizes (9.6). In this construction, certain node and arc numbers $\pi_{p+1}(x)$, $\gamma_{p+1}(x, y)$ are produced. The key to proving that f_{p+1} generates a temporally repeated flow that is maximal in $G(p)$ lies in these node and arc numbers, since they can be used to single out a cut in $G(p)$ that has capacity equal to $v(p)$ given by (9.6), thus proving that the flow is maximal and the cut minimal.

We proceed to a formal proof.

Decompose the flow f_{p+1} into a collection of chain flows from s to t. Let

$$(9.8) \qquad (x_1, x_2, \ldots, x_k; h), \qquad x_1 = s, \, x_k = t, \, h > 0,$$

be any one of the chain flows in this decomposition, and define correspondents of this chain flow in $G(p)$, namely

$$(9.9) \qquad (x_1(\tau_1), x_2(\tau_2), \ldots, x_k(\tau_k); h).$$

Here

$$(9.10) \qquad \tau_{i+1} = \tau_i + a(x_i, x_{i+1})$$

and

$$(9.11) \qquad \tau_i \geqslant 0, \qquad \tau_k \leqslant p.$$

This is to be done for all chain flows in the decomposition of f_{p+1}. (Although a chain decomposition of a flow f is not necessarily unique, this need cause no concern. Any decomposition will serve.)

That such chains exist in $G(p)$ follows from the second part of Theorem 3.6 by taking $\tau_1 = 0$. Then, since τ_k is the traversal time of the chain (9.8), we have

$$(9.12) \qquad \tau_k + \sum_{i=1}^{k-1} \gamma_{p+1}(x_i, x_{i+1}) = p + 1,$$

and since some arc number in this sum is positive (by virtue of the condition $h > 0$), it follows that $\tau_k \leqslant p$.

Hence, the number of correspondents (9.9) of (9.8) is given by

$$(9.13) \qquad p + 1 - \tau_k = \sum_{i=1}^{k-1} \gamma_{p+1}(x_i, x_{i+1}) > 0.$$

It is readily checked that the temporally repeated flow equal to the sum of all of the chain flows thus defined in $G(p)$ is really a flow in $G(p)$ from sources to sinks; the only thing remaining to be verified is that arc capacities in $G(p)$ are not violated. But this follows at once from the fact that f_{p+1} violates no arc capacity in G. From (9.13), the value $v(p)$ of this temporally repeated flow is

$$(9.14) \qquad v(p) = \sum_r (p + 1 - \sigma_r) h_r.$$

Here σ_r is the traversal time of the rth chain in the decomposition of f_{p+1} and h_r is the amount of flow along this chain. It follows that

$$(9.15) \qquad v(p) \geqslant (p + 1) v_{p+1} - \sum_{\mathscr{A}} a(x, y) f_{p+1}(x, y).$$

Here v_{p+1} is the value of f_{p+1}. By (9.15) and Corollary 3.2, we have

$$(9.16) \qquad v(p) \geqslant \sum_{\mathscr{A}} \gamma_{p+1}(x, y) c(x, y).$$

Now define the following set of arcs in $G(p)$:

$$(9.17) \quad \mathscr{C} = \{[x(\tau), y(\tau + a(x, y))] | \pi_{p+1}(x) \leqslant \tau < \pi_{p+1}(x) - a(x, y)\}.$$

In other words, $\mathscr{C}$ is the set of arcs that lead from any node of

$$(9.18) \qquad X = \{x(\tau) | \pi_{p+1}(x) \leqslant \tau\}$$

to its complement $\overline{X}$. Since every source of $G(p)$ is in X (because $\pi_{p+1}(s) = 0$) and every sink is in $\overline{X}$ (because $\pi_{p+1}(t) = p + 1$), it follows that $\mathscr{C} = (X, \overline{X})$ is a cut in $G(p)$. (Notice that this cut contains no hold-over arcs.) But from (9.17) and the definition of the arc numbers, the capacity of $\mathscr{C}$ is equal to

$$(9.19) \qquad \sum_{\mathscr{A}} \gamma_{p+1}(x, y) c(x, y).$$

Hence from (9.16), the temporally repeated flow generated by f_{p+1} is a maximal dynamic flow in $G(p)$, and the cut (9.17) is minimal. This proves

THEOREM 9.1. *The static flow f_{p+1} generates a temporally repeated dynamic flow that is maximal over all dynamic flows for p periods. This dynamic flow has value $v(p) = (p + 1) v_{p+1} - \sum a(x, y) f_{p+1}(x, y)$, where v_{p+1} is the value of f_{p+1}. The cut $\mathscr{C}$ defined by (9.17) is a minimal dynamic cut for p periods.*

A verbal way of describing the minimal dynamic cut $\mathscr{C}$ in terms of the arcs of the original network G is to say that the arc (x, y) of G first becomes a member of the cut at time $\tau = \pi_{p+1}(x)$ and remains in the cut for $\gamma_{p+1}(x, y)$ periods.

Since the routine of § 3 eventually stabilizes on a maximal static flow that minimizes total flow time $\sum a(x, y) f(x, y)$ over all maximal static flows, it follows that for all sufficiently large p, such a static flow generates maximal dynamic flows. Thus the maximal dynamic flow problem can be solved for all p by a finite (and efficient) process.

The following fact is worth mentioning. If $\bar{v}(p)$ denotes the maximal dynamic flow value for p periods, then we have, in view of Theorem 9.1 and Corollary 3.2,

$$(9.20) \qquad \bar{v}(p) - \bar{v}(p - 1) = v_{p+1}.$$

Thus, since the sequence v_p is monotone non-decreasing in p, so are successive differences of $\bar{v}(p)$. In other words, the piecewise linear curve obtained from the sequence of points $(p, \bar{v}(p))$, by joining adjacent points with line segments is convex.

We turn now to a different question concerning maximal dynamic flows, one that was raised and answered by Gale [32]. Consider a maximal dynamic flow for p periods, or equivalently, a maximal static flow in $G(p)$. What happens if we restrict this flow to $p' < p$ periods, that is, to $G(p')$. Will it still be maximal? The answer is no, in general. For example, a temporally repeated flow generated by f_{p+1} is maximal in $G(p)$, but may not be in $G(p')$. Even more, it can be seen that if there were a maximal dynamic flow for p periods whose restrictions are also maximal for all fewer periods, one may be forced to look outside the class of temporally repeated flows. The example of Fig. 9.1 is a case in point, since for $p = 5$, there is no temporally repeated flow in $G(5)$ that sends 1 unit into $t(3)$, 3 more units into $t(4)$, and 4 more units into $t(5)$, as does the flow of Fig. 9.7. It is true, nonetheless, that such "universal" maximal dynamic flows always exist.

An easy proof of this can be given from the second version of the supply-demand theorem (Corollary II.1.2) by setting up the demand schedule in $G(p)$:

$$(9.21) \qquad \begin{cases} \bar{v}(0) & \text{at } t(0), \\ \bar{v}(1) - \bar{v}(0) & \text{at } t(1), \\ \bar{v}(2) - \bar{v}(1) & \text{at } t(2), \\ \quad \cdots \\ \bar{v}(p) - \bar{v}(p - 1) & \text{at } t(p). \end{cases}$$

Here $\bar{v}(\tau)$ is the value of a maximal dynamic flow in $G(\tau)$, which can be assumed to be from $s(0)$ to $t(\tau)$. Now let X be an arbitrary subset of the sinks $t(\tau)$, $\tau = 0, 1, \ldots, p$, and let k be the largest value of τ for which $t(\tau)$ is in X. Then the aggregate demand over X does not exceed

$$\bar{v}(k) = \bar{v}(0) + \sum_{\tau=1}^{k} [\bar{v}(\tau) - \bar{v}(\tau - 1)].$$

But there is a flow from $s(0)$ to $t(k)$ in $G(p)$ that has value $\bar{v}(k)$. Hence, by the supply-demand theorem, the demands (9.21) are feasible.

It should be observed that this result makes no use of the simplifying assumptions that arc capacities and traversal times are independent of time, but rather holds for the more general case where these quantities change with time.

10. Project cost curves

A problem of some practical importance that has been discussed by Kelley and Walker [53] and Kelley [52] involves computing the cost curve for a "project" composed of many individual "jobs" or "activities." Here a project is a partially ordered set of jobs, the partial ordering arising from technological restrictions that force certain jobs to be finished before others can be started. It is assumed that each job has an associated normal completion time and a crash completion time, and that the cost of doing the job varies linearly between these two extreme times. Then it would be desirable to calculate the least project cost, given that the entire project must be completed in a prescribed time interval. This would yield one point on the project cost curve. Solving the problem for all feasible time intervals produces the complete project cost curve. With this information at hand, the project planner can answer either the question posed above, or the related question: given a fixed budget, what is the earliest project completion date?

We shall show how the project cost curve can be easily computed using network flows [29, 52].

There are at least two alternate ways of depicting the project as a directed network. For example, suppose the project consists of jobs 1, 2, 3, 4, 5 and that the only order relations are:

> 1   precedes   3, 4,
>
> 2   precedes   4,
>
> 3, 4   precede   5,

and those implied by transitivity. A usual way of picturing this partially ordered set is shown in Fig. 10.1, where nodes correspond to jobs and directed arcs to the displayed order relations. Another way is shown in Fig. 10.2, where some of the arcs represent jobs, and the nodes may be thought

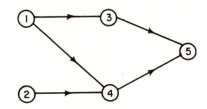

Figure 10.1

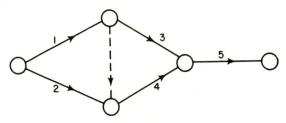

Figure 10.2

of as events in time; the existence of a node stipulates that all inward pointing jobs at the node must be completed before any outward pointing job can be started. Notice that the second of these two representations of the project uses an arc (the dotted one of Fig. 10.2) not corresponding to any job. This need cause no concern, since a dummy job can be added to the project to correspond to such an arc, and the assumption made that dummy jobs have zero completion time and zero cost. It is not difficult to see that allowing dummy jobs permits such a network representation for any project. Indeed, one could merely take the kind of network shown in Fig. 10.1, replace each node x by a pair of nodes x', x'' and add arcs (x', x'') to the network. Correspondents (x'', y') of the original arcs (x, y) then become dummies. But this is not, in general, efficient in terms of the number of nodes and arcs.

Using either of these network representations of the project, the problem of computing the cost curve can be shown to be a flow problem. We shall assume that a latter representation is at hand. Thus we take as given a directed network in which arcs correspond to jobs and nodes to events. This network contains no directed cycles. We may also assume, by adding beginning and terminal nodes s, t, if necessary, together with appropriate arcs pointing out from s and into t, that each arc is contained in some directed chain from s to t. We suppose that each arc (x, y) has associated with it three non-negative integers: $a(x, y)$, $b(x, y)$, $c(x, y)$, with $a(x, y) \leqslant b(x, y)$, the interpretation being that $a(x, y)$ is the crash time for

152

(x, y), $b(x, y)$ the normal completion time, while $c(x, y)$ is the decrease in cost of doing job (x, y) per unit increase in time from $a(x, y)$ to $b(x, y)$. In other words, the cost of doing (x, y) in $\tau(x, y)$ units of time is given by the known linear function

$$(10.1) \qquad k(x, y) - c(x, y)\tau(x, y)$$

over the interval

$$(10.2) \qquad a(x, y) \leqslant \tau(x, y) \leqslant b(x, y).$$

Then, given λ units of time in which to finish the project, the problem is to choose, for each job (x, y), a time $\tau(x, y)$ satisfying (10.2) in such a way that the resulting project cost

$$(10.3) \qquad \sum_{x,y} [k(x, y) - c(x, y)\tau(x, y)]$$

is minimized; or equivalently, the function

$$(10.4) \qquad \sum_{x,y} c(x, y)\tau(x, y)$$

is maximized. Thus, letting $\tau(x)$ be the (unknown) time of occurrence of event x, we wish to maximize (10.4) subject to the inequalities

$$(10.5) \qquad \tau(x, y) + \tau(x) - \tau(y) \leqslant 0,$$

$$(10.6) \qquad -\tau(s) + \tau(t) \leqslant \lambda,$$

$$(10.7) \qquad \tau(x, y) \leqslant b(x, y),$$

$$(10.8) \qquad -\tau(x, y) \leqslant -a(x, y).$$

Then the project cost $P(\lambda)$ corresponding to the assigned value of λ in (10.6) is given by

$$(10.9) \qquad P(\lambda) = \sum_{x,y} k(x, y) - \max \sum_{x,y} c(x, y)\tau(x, y),$$

the maximum being taken over all $\tau(x, y)$, $\tau(x)$ that satisfy the constraints. Here we assume the constraints are feasible, which will certainly be the case for large λ. Indeed, for given $\tau(x, y)$ satisfying (10.7) and (10.8), the constraints are feasible if and only if λ is at least equal to the τ-length of a longest chain from s to t. The proof of this relies on the fact that the project network contains no directed cycles.

Dummy jobs can be assumed to have lower bounds $a(x, y) = 0$, upper bounds $b(x, y) = 0$, and costs $c(x, y) = 0$ in this program.

It may be observed preliminarily that $P(\lambda)$, which is well defined for some λ-interval, is convex. For if λ_1, λ_2 are two given values of λ that make the constraints feasible, and if $\tau_1(x, y)$, $\tau_1(x)$, $\tau_2(x, y)$, $\tau_2(x)$ represent optimal solutions to the two corresponding programs, then averaging these two solutions gives a feasible solution to the constraints corresponding to the λ-value $\frac{1}{2}(\lambda_1 + \lambda_2)$. Hence, since we are minimizing $P(\lambda)$,

$$P\left(\frac{\lambda_1 + \lambda_2}{2}\right) \leqslant \tfrac{1}{2}P(\lambda_1) + \tfrac{1}{2}P(\lambda_2).$$

In addition, $P(\lambda)$ is piecewise linear, as will be apparent later on.

We may set $\tau(s) = 0$, since adding a constant to all event times does not alter the program. With this normalization, it follows from (10.5) that all $\tau(x)$ are non-negative, since the job times are non-negative by (10.8), and since each node is contained in some directed chain from s to t.

Let us examine the dual of the project cost program. If we assign non-negative multipliers $f(x, y)$, v, $g(x, y)$, $h(x, y)$ to the constraints (10.5), (10.6), (10.7), (10.8), respectively, the dual of the program, for fixed λ and $\tau(s) = 0$, has constraints

$$(10.10) \qquad f(x, y) + g(x, y) - h(x, y) = c(x, y),$$

$$(10.11) \qquad \sum_y [f(x, y) - f(y, x)] = \begin{cases} 0, & x \neq s, t \\ -v, & x = t, \end{cases}$$

subject to which

$$(10.12) \qquad \lambda v + \sum_{x,y} b(x, y)g(x, y) - \sum_{x,y} a(x, y)h(x, y)$$

is to be minimized. Here, we repeat, all variables are non-negative. Equalities appear in the constraints since variables of the primal program are not explicitly restricted in sign.

It follows immediately that at least one of $g(x, y)$, $h(x, y)$ can be taken zero in an optimal dual solution, and hence we may assume

$$(10.13) \qquad g(x, y) = \max\,[0, c(x, y) - f(x, y)],$$

$$(10.14) \qquad h(x, y) = \max\,[0, f(x, y) - c(x, y)].$$

Thus the dual problem becomes: find non-negative numbers $f(x, y)$, one for each arc of the project network, and a non-negative number v, that satisfy the flow equations (10.11) and minimize the non-linear function

$$(10.15) \qquad \lambda v + \sum_{x,y} b(x, y) \max\,[0, c(x, y) - f(x, y)]$$

$$- \sum_{x,y} a(x, y) \max\,[0, f(x, y) - c(x, y)].$$

The key observation at this point is that a function of f of the form

$$(10.16) \qquad b \max\,(0, c - f) - a \max\,(0, f - c)$$

(sketched in Fig. 10.3) is convex, and of course, piecewise linear. The convexity of (10.16) follows from the assumption $a \leqslant b$. Thus, even though

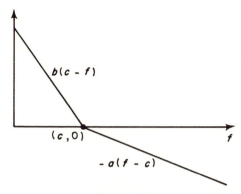

Figure 10.3

(10.15) is non-linear, it is the next best thing (for minimizing), namely, a sum of piecewise linear, convex functions of the individual variables. It is a well-known fact in linear programming theory that such a function can be dealt with by linear methods.†

Here one replaces each $f(x, y)$ by a sum of two non-negative variables, say

(10.17) $$f(x, y) = f(x, y; 1) + f(x, y; 2),$$

the new variables being subject to the capacity constraints

(10.18) $$f(x, y; 1) \leqslant c(x,y),$$

(10.19) $$f(x, y; 2) \leqslant \infty.$$

Then $f(x, y; 1)$ has coefficient $-b(x, y)$, $f(x, y; 2)$ has coefficient $-a(x, y)$ in the new minimizing form. Thus, if we define

(10.20) $$c(x, y; k) = \begin{cases} c(x, y), & k = 1, \\ \infty, & k = 2, \end{cases}$$

---

† The idea is simply this. Replace each variable by a sum of bounded non-negative variables, where each variable in the sum corresponds to one of the pieces of the cost function for the original variable. Make up a new, linear cost function by assigning each of the new variables a cost coefficient equal to the slope of its linear piece of the original cost form. Thus, for example, if the original cost function for the non-negative variable x of the program has breakpoints at $0 < b_1 < b_2 < \ldots < b_k$, one makes the replacement $x = x_1 + x_2 + \ldots + x_{k+1}$, where $0 \leqslant x_1 \leqslant b_1$, $0 \leqslant x_2 \leqslant b_2 - b_1, \ldots, 0 \leqslant x_k \leqslant b_k - b_{k-1}$, $0 \leqslant x_{k+1} \leqslant \infty - b_k = \infty$. Since the costs of the new variables are monotone increasing, it follows that in a minimizing solution, if some $x_i > 0$, then all preceding x_j are at their upper bounds. Hence the replacement is legitimate.

(10.21)
$$a(x, y; k) = \begin{cases} b(x, y), & k = 1, \\ a(x, y), & k = 2, \end{cases}$$

the dual program has constraints

(10.22)
$$\sum_{y,k} [f(x, y; k) - f(y, x; k)] = \begin{cases} 0, & x \neq s, t, \\ -v, & x = t, \end{cases}$$

(10.23)
$$0 \leqslant f(x, y; k) \leqslant c(x, y; k),$$

and minimizing form

(10.24)
$$\lambda v - \sum_{x,y,k} a(x, y; k) f(x, y; k).$$

This program has the following flow interpretation. First enlarge the project network by doubling the number of arcs: corresponding to each arc (x, y) of the project network there are now two arcs $(x, y; 1)$ and $(x, y; 2)$ from x to y (see Fig. 10.4). Each arc $(x, y; k)$ of the new network has an assigned capacity $c(x, y; k)$. The problem is to construct a flow from s to t of value v in the new network that minimizes (10.24).

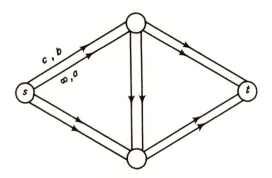

Figure 10.4

Except for the details that minimization has replaced maximization in (10.24), and pairs of arcs join nodes, the problem is now in familiar form. The second of these minor differences could be eliminated, if desired, by inserting an additional node in the "middle" of one arc of each pair. However, this greatly increases the number of nodes and arcs needlessly. A better way to handle multiple arcs joining nodes is simply to augment the information contained in the labels assigned to nodes during the labeling process by indicating which arc produced the label. We do this in the algorithm, outlined below, which is also designed to minimize (10.24).

This algorithm, like the one of § 3, solves the problem for all λ and thus generates the complete project cost curve. One other slight variation which deserves preliminary mention is that, before performing the usual labeling process, a check is made (using a labeling process) to see whether "infinite breakthrough" is possible, that is, whether there is a chain of admissible arcs from s to t, each of which has infinite capacity. For the existence of such a chain means that further decrease in λ would make the constraints (10.5)–(10.8) infeasible, and this signals termination of the computation.

We use the notation $\tau(x)$ for node numbers in the algorithm, instead of our usual $\pi(x)$, because these node numbers do indeed have the interpretation of event times in the original program. The algorithm begins with the zero flow and an assignment of node numbers produced by finding a chain of maximal b-length from s to t. Thus $\tau(s) = 0$ and $\tau(t)$ equals the length of this chain. Then $\lambda = \tau(t)$ is the largest λ of interest, since the project can be completed in λ time units even if all job times are at their upper bounds. The node numbers partition the arcs into admissible and inadmissible classes in the usual way, and the labeling process (modified as mentioned above) is then performed on admissible arcs. Following non-breakthrough, the node numbers (event times) are changed by subtracting a positive integer from those corresponding to unlabeled nodes. This produces a smaller value of λ, namely the new $\tau(t)$, and consequently another point on the project cost curve $P(\lambda)$. Moreover, optimal job times $\tau(x, y)$ corresponding to $\lambda = \tau(t)$ are given simply by defining

$$(10.25) \qquad \tau(x, y) = \min\,[b(x, y), \tau(y) - \tau(x)].$$

We shall discuss these assertions in more detail following the algorithm statement.

That the project network contains no directed cycles comes into play in starting the algorithm for the dual flow problem, since this means that the method of § 5 can be used to find an initial assignment of node numbers. Indeed, were it not for the absence of directed cycles, the constraints (10.5)–(10.8) would be infeasible in general, as is easily seen by summing (10.5) around a directed cycle. This is also reflected in the dual problem, as the form (10.24) would be unbounded on its constraint set if the project network contained a directed cycle. For an infinite amount of flow could be sent around this cycle without changing v, so that (10.24) would, in general, be negatively infinite.

Optimality properties for the program (10.22)–(10.24) are

$$(10.26) \qquad\qquad \tau(s) = 0, \qquad \tau(t) = \lambda,$$

$$(10.27) \qquad a(x, y; k) + \tau(x) - \tau(y) < 0 \Rightarrow f(x, y; k) = 0,$$

$$(10.28) \quad a(x, y; k) + \tau(x) - \tau(y) > 0 \Rightarrow f(x, y; k) = c(x, y; k).$$

Here (10.27) and (10.28) are just the reverse of (3.19) and (3.20), as is to be expected. The algorithm below produces successive flows and node numbers satisfying these properties for decreasing values of λ.

To shorten the notation, we set

$$(10.29) \qquad \bar{a}(x, y; k) = a(x, y; k) + \tau(x) - \tau(y).$$

Arcs for which $\bar{a}(x, y; k) = 0$ are admissible.

Start. (Finding a chain of maximal b-length.) Use the shortest chain algorithm of § 5, where each arc (x, y) of the project network is assigned the length $-b(x, y)$. At the conclusion of this routine, (negative) node numbers $\pi(x)$ will have been generated, with $\pi(s) = 0$. Set $\tau(x) = -\pi(x)$. Take all $f(x, y; k) = 0$. (The properties (10.26)–(10.28) now hold for $\lambda = \tau(t)$.)

Iterative procedure. Enter with an integral flow $f(x, y; k)$ and node integers $\tau(x)$ satisfying (10.26)–(10.28) for some λ. (During the iterative procedure, a label assigned to node y will be of the form $[x, k^{\pm}, \varepsilon(y)]$. Here x is a node, k^+ indicates that the arc $(x, y; k)$ was used to label y from x; k^- that the arc $(y, x; k)$ was used to label y from x; and $\varepsilon(y)$ indicates the largest flow change along the path from s to y.)

First labeling. Start by labeling s with $[-, -, \varepsilon(s) = \infty]$. The only labeling rule is: node y can be labeled from (labeled) node x if $(x, y; 2)$ is admissible; y then receives the label $[x, 2^+, \varepsilon(y) = \infty]$. If breakthrough, terminate. If non-breakthrough, go on to the second labeling.

Second labeling. Nodes labeled above retain their labels, and the labeling process continues as follows. All nodes revert to the unscanned state. When scanning a labeled node x, the labeling rules are: y can be labeled from x if either

    (a) $(x, y; k)$ is admissible and $f(x, y; k) < c(x, y; k)$,

    (b) $(y, x; k)$ is admissible and $f(y, x; k) > 0$.

In case (a), y receives the label $[x, k^+, \varepsilon(y)]$ where $\varepsilon(y) = \min [\varepsilon(x), c(x, y; k) - f(x, y; k)]$; in case (b), y receives the label $[x, k^-, \varepsilon(y)]$, where $\varepsilon(y) = \min [\varepsilon(x), f(y, x; k)]$. If breakthrough, change the flow by adding and subtracting $\varepsilon(t)$ along the path from s to t picked out by the labels. If non-breakthrough, single out the following subsets of arcs:

$$\mathscr{A}_1 = \{(x, y; k) | x \text{ labeled, } y \text{ unlabeled, } \bar{a}(x, y; k) < 0\},$$

$$\mathscr{A}_2 = \{(x, y; k) | x \text{ unlabeled, } y \text{ labeled, } \bar{a}(x, y; k) > 0\},$$

and define

$$\delta_1 = \min_{\mathscr{A}_1} [-\bar{a}(x, y; k)],$$

$$\delta_2 = \min_{\mathscr{A}_2} [\bar{a}(x, y; k)],$$

$$\delta = \min (\delta_1, \delta_2).$$

Change the node numbers $\tau(x)$ by subtracting δ from all $\tau(x)$ corresponding to unlabeled x. Discard the old labels and repeat.

That the algorithm produces successive flows and node numbers satisfying the optimality properties (10.26), (10.27), (10.28) is readily checked, just as in the mininal cost flow routine of § 3. It is also easy to see that termination occurs; that is, at some stage, the first labeling results in (infinite) breakthrough. For, suppose that the algorithm fails to terminate, so that an infinite sequence of finite breakthroughs and non-breakthroughs occurs. The number of breakthroughs in this sequence is finite. For otherwise, flows having arbitrarily large values v would be produced. But such a flow must contain a chain flow along admissible arcs corresponding to $k = 2$ (the infinite capacity arcs). Hence at some stage there is a chain from s to t of admissible arcs corresponding to $k = 2$, and thus the first labeling would produce breakthrough. This leaves only the possibility that infinitely many successive non-breakthroughs occur. This possibility is eliminated, just as in § 3, by the fact that at least one more node can always be labeled following non-breakthrough.

It may also be checked that the sets $\mathscr{A}_1$, $\mathscr{A}_2$ that define the node number change δ cannot both be empty. (In fact, $\mathscr{A}_1$ cannot be empty.) For if both $\mathscr{A}_1$ and $\mathscr{A}_2$ are empty, the flow at that stage is maximal, hence has infinite value. But this is absurd, as termination would have occurred. Thus δ is a positive integer.

As was remarked earlier, each new set of event times $\tau(x)$ yields a new point on the project cost curve by defining $\tau(x, y)$ as in (10.25) and calculating

$$P(\lambda) = P[\tau(y)] = \sum_{x,y} [k(x, y) - c(x, y)\tau(x, y)].$$

We now verify that (10.25) does define optimal job times corresponding to $\lambda = \tau(t)$. To this end, we go back to the original pair of dual programs (10.4)–(10.8) and (10.10)–(10.12), using also (10.13), (10.14) to define g and h, and (10.17) to define f. It suffices to show that

(10.30) $\qquad \tau(x, y) + \tau(x) - \tau(y) < 0 \Rightarrow f(x, y) = 0,$

(10.31) $\qquad \tau(x, y) < b(x, y) \Rightarrow g(x, y) = 0,$

(10.32) $\qquad \tau(x, y) > a(x, y) \Rightarrow h(x, y) = 0,$

since (with $\tau(s) = 0$, $\tau(t) = \lambda$) these are optimality properties for primal and dual. If the hypothesis of (10.30) holds, then $\tau(x, y) = b(x, y)$, hence $b(x, y) + \tau(x) - \tau(y) < 0$. Consequently $a(x, y) + \tau(x) - \tau(y) < 0$ also. It then follows from (10.27) that $f(x, y; k) = 0$, $k = 1$ and 2, hence $f(x, y) = f(x, y; 1) + f(x, y; 2) = 0$, verifying (10.30). Suppose next that

159

$\tau(x, y) < b(x, y)$. Then $\tau(x, y) = \tau(y) - \tau(x) < b(x, y)$, hence by (10.28), $f(x, y; 1) = c(x, y)$. Then $f(x, y) \geqslant c(x, y)$, hence $g(x, y) = \max[0, c(x, y) - f(x, y)] = 0$, proving (10.31). Finally, assume $\tau(x, y) > a(x, y)$. If $\tau(x, y) = \tau(y) - \tau(x)$, then $a(x, y) + \tau(x) - \tau(y) < 0$, hence by (10.27), $f(x, y; 2) = 0$. Consequently $f(x, y) \leqslant c(x, y)$, so that $h(x, y) = \max[0, f(x, y) - c(x, y)] = 0$. If, on the other hand, $\tau(x, y) = b(x, y) < \tau(y) - \tau(x)$, then $a(x, y) < \tau(y) - \tau(x)$, and again we deduce $h(x, y) = 0$. This completes the proof that $\tau(x, y)$ defined by (10.25), together with the event times $\tau(x)$, solves the original project cost program.

The function $P(\lambda)$ is linear between successive values of $\lambda = \tau(t)$ generated by the algorithm. We shall show how to pick out the breakpoints of the convex, piecewise linear $P(\lambda)$. Not every value of $\tau(t)$ is necessarily a breakpoint.

Suppose that $\lambda_1 > \lambda_2$ are two successive λ's, and let λ satisfy $\lambda_1 \geqslant \lambda \geqslant \lambda_2$. Let f be the flow that produced the node number change yielding λ_2 from λ_1, and suppose f has value v. Since f minimizes (10.24) for $\lambda = \lambda_1$, it follows that

$$(10.33) \qquad P(\lambda_1) = K - (\lambda_1 v - \sum_{x,y,k} a(x, y; k) f(x, y; k)).$$

Here K is the constant $\sum_{x,y} [k(x, y) - b(x, y)c(x, y)]$. But f also minimizes (10.24) for any λ in the interval $\lambda_1 \geqslant \lambda \geqslant \lambda_2$. (This is analogous to Corollary 3.4.) Hence

$$(10.34) \qquad P(\lambda) = K - \left(\lambda v - \sum_{x,y,k} a(x, y; k) f(x, y; k)\right),$$

and consequently

$$(10.35) \qquad\qquad P(\lambda) - P(\lambda_1) = (\lambda_1 - \lambda)v, \qquad\qquad \lambda_1 \geqslant \lambda \geqslant \lambda_2.$$

Thus $P(\lambda)$ is linear between successive values of $\tau(t)$. Now suppose $\lambda_1 > \lambda_2 > \lambda_3$ are three successive values of $\tau(t)$ generated in the computation, and let v be as defined above. Suppose also that v' is the value of the flow that produced the non-breakthrough yielding λ_3. Then

$$P(\lambda_2) - P(\lambda_1) = (\lambda_1 - \lambda_2)v,$$
$$P(\lambda_3) - P(\lambda_2) = (\lambda_2 - \lambda_3)v'.$$

Consequently λ_2 is a breakpoint of $P(\lambda)$ if and only if $v < v'$; that is, if and only if there is an intervening breakthrough between the two non-breakthroughs that yield λ_2 and λ_3.

For example, if a problem computation results in the sequence of breakthroughs and non-breakthroughs (indicated by B and N)

$$B \; \textcircled{N} \; B \; B \; N \; \textcircled{N} \; B \; N \; N \; \textcircled{N} \; B,$$

then the circled non-breakthroughs suffice to define $P(\lambda)$.

At the conclusion of the computation, a chain of admissible links corresponding to $k = 2$ has been located. Summing the equalities $a(x, y) + \tau(x) = \tau(y)$ along this chain shows that $\lambda = \tau(t)$ is equal to the a-length of this chain. Consequently the project cannot be completed in any smaller time interval.

The method of this section can also be used to compute project cost curves in case the given job costs are assumed piecewise linear and convex between crash and normal completion times $a(x, y)$ and $b(x, y)$. This merely introduces more arcs from x to y into the flow network; in fact, one more arc for each additional breakpoint of the function giving the cost of job (x, y).

We conclude this section with Table 10.1 summarizing the solution of the numerical example shown in Fig. 10.5. Note the behavior of the optimal

$\tau(1)$	0	0	0	0	0	0	0	—
$\tau(2)$	3	3	3	3	2	2	1	—
$\tau(3)$	5	4	3	3	2	2	2	—
$\tau(4) = \lambda$	11	10	9	8	7	4	3	—
$\tau(1, 2)$	3	3	3	3	2	2	1	—
$\tau(1, 3)$	4	4	3	3	2	2	2	—
$\tau(2, 3)$	2	1	0	0	0	0	1	—
$\tau(2, 4)$	5	5	5	5	5	2	2	—
$\tau(3, 4)$	6	6	6	5	5	2	1	—
$f(1, 2)$	0	1	1	2	3	3	3	∞
$f(1, 3)$	0	0	1	1	1	1	2	∞
$f(2, 3)$	0	1	1	2	2	2	1	1
$f(2, 4)$	0	0	0	0	1	1	2	∞
$f(3, 4)$	0	1	2	3	3	3	3	∞
v	0	1	2	3	4	4	5	∞
$P(\lambda) - K$	0	1	3	6	10	22	27	—

Table 10.1

job times $\tau(2, 3)$ as λ varies from 11 to 3 in this example. For $\lambda = 4$, it is optimal to take $\tau(2, 3)$ at its lower bound, but further decrease in λ implies an increase in $\tau(2, 3)$ away from its lower bound. This kind of behavior

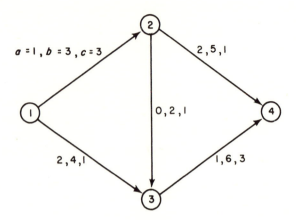

Figure 10.5

may go against one's intuition at first, but a little reflection shows that it is not, after all, surprising.

11. Constructing minimal cost circulations [28]

The method presented here for computing optimal network flows is more general than those described earlier in at least three ways:

    (a) lower bounds as well as capacities are assumed for each arc flow, and are dealt with directly;

    (b) the cost coefficient for an arc is arbitrary in sign;

    (c) the method can be initiated with any circulation, feasible or not, and any set of node numbers.

(It is convenient to describe the computation in terms of circulations, rather than flows from sources to sinks.) The freedom to begin with any circulation and node numbers, instead of starting with particular ones which satisfy certain optimality properties, as has been the case before, is perhaps the most important practical feature of the method. For example, in actual applications, one is often interested in seeing what changes will occur in an optimal solution when some of the given data are altered. This method is tailored for such an examination, since the old optimal primal and dual solutions can be used to start the new problem, thereby greatly decreasing computation time.

An interesting feature of the method is that, loosely speaking, the status of no arc of the network is worsened at any step of the computation. We shall make this statement more precise later on.

162

We take the problem in circulation form. That is, we want to construct f that satisfies

(11.1) $$f(x, N) - f(N, x) = 0, \qquad \text{all } x \in N,$$

(11.2) $$l(x, y) \leqslant f(x, y) \leqslant c(x, y), \qquad \text{all } (x, y) \in \mathscr{A};$$

and minimizes the linear cost function

(11.3) $$\sum_{\mathscr{A}} a(x, y) f(x, y).$$

(Here $0 \leqslant l(x, y) \leqslant c(x, y)$, and as usual, we assume integral data.) Thus, if it is desired to construct a feasible flow from s to t of given value v that minimizes (11.3), one can merely add a return flow arc (t, s) with $l(t, s) = c(t, s) = v, a(t, s) = 0$, to get the problem in circulation form. Or, if it is desired to construct a maximal feasible flow from s to t that minimizes (11.3), one can take $l(t, s) = 0$, $c(t, s)$ large, $a(t, s)$ negatively large.

Of course feasible circulations may not exist. In this case the algorithm terminates with the location of a subset X of nodes for which the condition of Theorem II.3.1 is violated.

For given node numbers π, we set

(11.4) $$\bar{a}(x, y) = a(x, y) + \pi(x) - \pi(y).$$

Then, for given π and circulation f, an arc (x, y) is in just one of the following states:

$$(\alpha) \ \bar{a}(x, y) > 0, \ f(x, y) = l(x, y),$$
$$(\beta) \ \bar{a}(x, y) = 0, \ l(x, y) \leqslant f(x, y) \leqslant c(x, y),$$
$$(\gamma) \ \bar{a}(x, y) < 0, \ f(x, y) = c(x, y),$$
$$(\alpha_1) \ \bar{a}(x, y) > 0, \ f(x, y) < l(x, y),$$
$$(\beta_1) \ \bar{a}(x, y) = 0, \ f(x, y) < l(x, y),$$
$$(\gamma_1) \ \bar{a}(x, y) < 0, \ f(x, y) < c(x, y),$$
$$(\alpha_2) \ \bar{a}(x, y) > 0, \ f(x, y) > l(x, y),$$
$$(\beta_2) \ \bar{a}(x, y) = 0, \ f(x, y) > c(x, y),$$
$$(\gamma_2) \ \bar{a}(x, y) < 0, \ f(x, y) > c(x, y).$$

We say that an arc is *in kilter* if it is in one of the states α, β, γ; otherwise the arc is *out of kilter*. Thus to solve the problem, it suffices to get all arcs in kilter, since optimality properties are

(11.5) $$\bar{a}(x, y) < 0 \Rightarrow f(x, y) = c(x, y),$$

(11.6) $$\bar{a}(x, y) > 0 \Rightarrow f(x, y) = l(x, y).$$

With each state that an arc (x, y) can be in, we associate a non-negative number, called the *kilter number* of the arc in the given state. An in-kilter arc has kilter number 0; the arc kilter numbers corresponding to each out-of-kilter state are listed below:

$$(\alpha_1) \text{ or } (\beta_1): l(x, y) - f(x, y),$$
$$(\gamma_1): \bar{a}(x, y)[f(x, y) - c(x, y)],$$
$$(\alpha_2): \bar{a}(x, y)[f(x, y) - l(x, y)],$$
$$(\beta_2) \text{ or } (\gamma_2): f(x, y) - c(x, y).$$

Thus out-of-kilter arcs have positive kilter numbers. The kilter numbers for states $\alpha_1, \beta_1, \beta_2, \gamma_2$ measure infeasibility for the arc flow $f(x, y)$, while the kilter numbers for states γ_1, α_2 are, in a sense, a measure of the degree to which the optimality properties (11.5), (11.6) fail to be satisfied.

The algorithm concentrates on a particular out-of-kilter arc and attempts to put it in kilter. It does this in such a way that all in-kilter arcs stay in kilter, whereas the kilter number for any out-of-kilter arc either decreases or stays the same. Thus all arc kilter numbers are monotone non-increasing throughout the computation. (This is the interesting feature of the method that was mentioned previously.) However, steps can occur that change no kilter number, and this somewhat complicates the proof of termination. But if the process begins with a feasible circulation, the monotone property is stronger: at least one arc kilter number decreases at each step, thus providing a simpler proof of finiteness in this case.

A basic notion underlying the method is to utilize the labeling process of II.3, modified appropriately, for increasing or decreasing a particular arc flow in a circulation. The appropriate modification this time will not be in terms of the notion of "admissibility" for an arc, used previously, but will rather be more general.

We now state the algorithm.

The out-of-kilter algorithm.† Enter with any integral circulation f and any set of node integers π. Next locate an out-of-kilter arc (s, t) and go on to the appropriate case below.

(α_1) $\bar{a}(s, t) > 0$, $f(s, t) < l(s, t)$. Start a labeling process at t, trying to reach s, first assigning t the label $[s^+, \varepsilon(t) = l(s, t) - f(s, t)]$. The labeling rules are:

(11.7) If x is labeled $[z^\pm, \varepsilon(x)]$, y is unlabeled, and if (x, y) is an arc such that either

(a) $\bar{a}(x, y) > 0$, $f(x, y) < l(x, y)$,

(b) $\bar{a}(x, y) \leqslant 0$, $f(x, y) < c(x, y)$,

---

† An IBM 704 code based on this algorithm has been prepared by J.D. Little. A FORTRAN-FAP revision for the IBM 7090 has been written by R. Clasen. This code, identified as RS OKF1, is available through SHARE. A sample problem involving 2900 arcs and 775 nodes required 1139 breakthroughs, 411 non-breakthroughs. Total computing time was 5 minutes, exclusive of an input-output time of 3.2 minutes.

then y receives the label $[x^+, \varepsilon(y)]$, where

$\varepsilon(y) = \min\,[\varepsilon(x), l(x, y) - f(x, y)]$ in case (a),

$\varepsilon(y) = \min\,[\varepsilon(x), c(x, y) - f(x, y)]$ in case (b).

(11.8) If x is labeled $[z^\pm, \varepsilon(x)]$, y is unlabeled, and if (y, x) is an arc such that either

(a) $\bar{a}(y, x) \geqslant 0, f(y, x) > l(y, x)$,

(b) $\bar{a}(y, x) < 0, f(y, x) > c(y, x)$,

then y receives the label $[x^-, \varepsilon(y)]$, where

$\varepsilon(y) = \min\,[\varepsilon(x), f(y, x) - l(y, x)]$ in case (a),

$\varepsilon(y) = \min\,[\varepsilon(x), f(y, x) - c(y, x)]$ in case (b).

If breakthrough occurs (that is, s receives a label), so that a path from t to s has been found, change the circulation f by adding $\varepsilon(s)$ to the flow in forward arcs of this path, subtracting $\varepsilon(s)$ from the flow in reverse arcs, and finally adding $\varepsilon(s)$ to $f(s, t)$. If non-breakthrough, let X and $\overline{X}$ denote labeled and unlabeled sets of nodes, and define two subsets of arcs:

(11.9) $\mathscr{A}_1 = \{(x, y)|x \in X, y \in \overline{X}, \bar{a}(x, y) > 0, f(x, y) \leqslant c(x, y)\}$,

(11.10) $\mathscr{A}_2 = \{(y, x)|x \in X, y \in \overline{X}, \bar{a}(y, x) < 0, f(y, x) \geqslant l(y, x)\}$.

Then let

(11.11) $$\delta_1 = \min_{\mathscr{A}_1}\,[\bar{a}(x, y)],$$

(11.12) $$\delta_2 = \min_{\mathscr{A}_2}\,[-\bar{a}(y, x)],$$

(11.13) $$\delta = \min\,(\delta_1, \delta_2).$$

(Here δ_i is a positive integer or ∞ according as $\mathscr{A}_i$ is non-empty or empty.) Change the node integers by adding δ to all $\pi(x)$ for $x \in \overline{X}$.

(β_1) or (γ_1). $\bar{a}(s,\ t) = 0,\ f(s, t) < l(s, t)$ or $\bar{a}(s, t) < 0,\ f(s, t) < c(s, t)$. Same as (α_1), except $\varepsilon(t) = c(s, t) - f(s, t)$.

(α_2) or (β_2). $\bar{a}(s, t) > 0,\ f(s, t) > l(s, t)$, or $\bar{a}(s, t) = 0,\ f(s, t) > c(s, t)$. Here the labeling process starts at s, in an attempt to reach t. Node s is assigned the label $[t^-, \varepsilon(s) = f(s, t) - l(s, t)]$. The labeling rules are (11.7) and (11.8) again. If breakthrough, change the circulation by adding and subtracting $\varepsilon(t)$ to arc flows along the path from s to t; then subtract $\varepsilon(t)$ from $f(s, t)$. If non-breakthrough, change the node numbers as above.

(γ_2). $\bar{a}(s, t) < 0,\ f(s, t) > c(s, t)$. Same as (α_2) or (β_2), except $\varepsilon(s) = f(s, t) - c(s, t)$.

The labeling process is repeated for the arc (s, t) until either (s, t) is in kilter, or until a non-breakthrough occurs for which $\delta = \infty$. In the latter case, stop. (There is no feasible circulation.) In the former case, locate another out-of-kilter arc and continue.

We show that the out-of-kilter algorithm terminates, and that all arc kilter numbers are monotone non-increasing throughout the computation.

Suppose that arc (s, t) is out of kilter, say in state α_1. The origin for labeling is t, the terminal s. The arc (s, t) cannot be used to label s directly since neither (11.8a) nor (11.8b) is applicable. Consequently, if break-through occurs, the resulting path from t to s, together with the arc (s, t), is a cycle. Then the flow changes that are made on arcs of this cycle again yield a circulation. Moreover, the labeling rules have been selected in such a way that kilter numbers for arcs of this cycle do not increase, and at least one, namely, for arc (s, t), decreases by a positive integer. Kilter num-bers for arcs not in the cycle of course do not change.

Similar remarks apply if (s, t) is in one of the other out-of-kilter states.

We summarize the possible effects of a breakthrough on an arc (x, y) in Fig. 11.1, which shows the state transitions that may occur following breakthrough. If a transition is possible, the number recorded beside the corresponding arrow represents the change in kilter number. (Here ε is the flow change.)

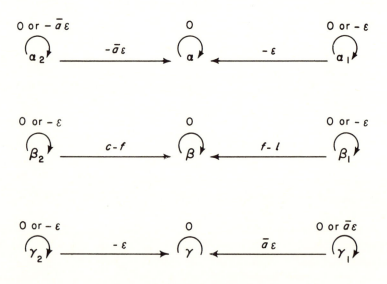

Figure 11.1   Breakthrough diagram

Verification of the breakthrough diagram is straightforward. For example, suppose arc (x, y) is in state α_2, with $\bar{a}(x, y) > 0, f(x, y) > l(x, y)$, and kilter number $\bar{a}(x, y) \, [f(x, y) - l(x, y)] > 0$. If (x, y) is not an arc of the cycle of flow changes, then (x, y) remains in state α_2 with zero change in kilter number. If the flow in arc (x, y) has changed as a result of the break-through, then either (x, y) is the arc (s, t) or, by the labeling rules, (x, y) is a reverse arc of the path from origin to terminal. Specifically, x was labeled from y using (11.8a). In either case, $f(x, y)$ decreases by the positive integer $\varepsilon \leqslant f(x, y) - l(x, y)$, the new state for (x, y) is α_2 or α, and hence the kilter number for (x, y) has decreased by $\varepsilon \bar{a}(x, y) > 0$. The rest of the diagram may be verified similarly.

The state transitions and changes in kilter number that may occur following a non-breakthrough with $\delta < \infty$ are indicated in Fig. 11.2.

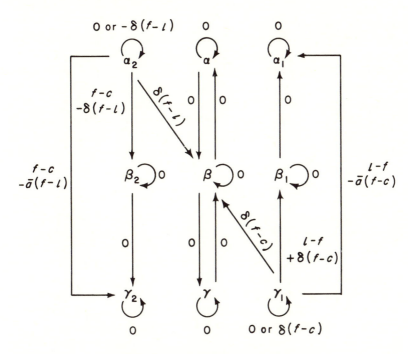

Figure 11.2   Non-breakthrough diagram

Again we omit a detailed verification, but consider, for example, an arc (x, y) in state γ_1, so that $\bar{a}(x, y) < 0$, $f(x, y) < c(x, y)$, having kilter number $\bar{a}(x, y)\,[f(x, y) - c(x, y)] > 0$ before the node number change is made. If both x and y are in X or both in $\bar{X}$, then $\bar{a}(x, y)$ remains the same after the node number change, and consequently (x, y) stays in state γ_1 with no change in kilter number. We cannot have x in X, y in $\bar{X}$ (labeling rule (11.7b)), and hence the remaining possibility is x in $\bar{X}$, y in X. Then $\bar{a}(x, y)$ is increased by $\delta > 0$. Consequently the arc (x, y) either remains in state γ_1 (if $\delta < -\bar{a}(x, y)$), goes into state β (if $\delta = -\bar{a}(x, y)$ and $f(x, y) \geqslant l(x, y)$), into state β_1 (if $\delta = -\bar{a}(x, y)$ and $f(x, y) < l(x, y)$), or into state

α_1 (if $\delta > -\bar{a}(x, y)$ and $f(x, y) < l(x, y)$), and the corresponding changes in kilter number are respectively

$$\delta[f(x, y) - c(x, y)] < 0,$$
$$\delta[f(x, y) - c(x, y)] < 0,$$
$$l(x, y) - f(x, y) + \delta[f(x, y) - c(x, y)] \leqslant 0,$$
$$l(x, y) - f(x, y) - \bar{a}(x, y)[f(x, y) - c(x, y)] \leqslant 0.$$

(The remaining logical possibility $\delta > -\bar{a}(x, y)$, $f(x, y) \geqslant l(x, y)$ cannot occur, since if $f(x, y) \geqslant l(x, y)$, then (x, y) is in $\mathscr{A}_2$ defined by (11.10) and hence $\delta \leqslant -\bar{a}(x, y)$.)

It follows from the breakthrough and non-breakthrough diagrams that kilter numbers are monotone non-increasing throughout the computation. Moreover, if breakthrough occurs, at least one arc kilter number decreases by a positive integer. Thus to establish termination, it suffices to show that an infinite sequence of consecutive non-breakthroughs, each with $\delta < \infty$, is impossible. To show this, let us suppose that a labeling resulting in non-breakthrough with $\delta < \infty$ has occurred, and let X, $\bar{X}$ denote labeled and unlabeled sets of nodes. After changing the node numbers, the new function $\bar{a}'(x, y)$ is given by

$$(11.14) \qquad \bar{a}'(x, y) = \begin{cases} \bar{a}(x, y) - \delta & \text{for } x \text{ in } X, y \text{ in } \bar{X}, \\ \bar{a}(x, y) + \delta & \text{for } x \text{ in } \bar{X}, y \text{ in } X, \\ \bar{a}(x, y) & \text{otherwise.} \end{cases}$$

If the arc (s, t) is still out of kilter, then the origin is the same for the next labeling, and it follows from (11.14) and the labeling rules that every node of X will again be labeled. Thus if the new labeling results in non-breakthrough with labeled set X', we have $X \subseteq X'$. Let $\mathscr{A}'_1$, $\mathscr{A}'_2$ denote the new sets defined in terms of X', $\bar{a}'$, and f by (11.9), (11.10), and suppose $X = X'$. Then from (11.14) we have $\mathscr{A}'_1 \subseteq \mathscr{A}_1$, $\mathscr{A}'_2 \subseteq \mathscr{A}_2$, and at least one of these inclusions is proper by (11.11), (11.12), (11.13). Hence the new labeling either assigns a label to at least one more node, or failing this, an arc is removed from one of the sets $\mathscr{A}_1$ or $\mathscr{A}_2$. It follows that, after finitely many non-breakthroughs with $\delta < \infty$, we either get the arc (s, t) in kilter, obtain a breakthrough, or obtain a non-breakthrough with $\delta = \infty$.

If a non-breakthrough with $\delta = \infty$ occurs, there is no feasible circulation. For if $\delta = \infty$, then from the definitions of $\mathscr{A}_1$, $\mathscr{A}_2$ and the labeling rules, we have $f(x, y) \geqslant c(x, y)$, $f(y, x) \leqslant l(y, x)$ for $x \in X$, $y \in \bar{X}$. Moreover, for the arc (s, t), either t is in X, s in $\bar{X}$ with $f(s, t) < l(s, t)$, or s is in X, t in $\bar{X}$ with $f(s, t) > c(s, t)$. (This is immediate for cases $\alpha_1, \beta_1, \beta_2, \gamma_2$ of the algorithm, and follows from (11.9) and the assumption $\delta = \infty$ for case α_2, from (11.10) and the assumption $\delta = \infty$ for case γ_1.) Hence, summing the conservation equations (11.1) over x in X, we obtain in all cases

$$0 = f(X, \overline{X}) - f(\overline{X}, X) > c(X, \overline{X}) - l(\overline{X}, X).$$

But this violates the feasibility condition of Theorem II.3.1. Thus $\delta = \infty$ implies there is no feasible circulation.

THEOREM 11.1. *The out-of-kilter algorithm either solves the problem* (11.1), (11.2), (11.3) *in finitely many applications of the labeling process or terminates with the conclusion that no feasible circulation exists. All arc kilter numbers are monotone non-increasing throughout the computation. In addition, if the algorithm is initiated with a feasible circulation, at least one arc kilter number decreases with each labeling.*

The only part of Theorem 11.1 that remains to be checked is the last assertion. If the computation begins with a feasible circulation, the states $\alpha_1, \beta_1, \beta_2, \gamma_2$ are empty to begin with, and consequently remain empty through the computation. Hence, at each non-breakthrough (as well as each breakthrough), the kilter number for at least one arc, namely (s, t), decreases by a positive integer.

It is worth while to point out how the out-of-kilter algorithm generalizes the method of § 3 for constructing a maximal flow from source s to sink t that minimizes cost over all maximal flows. Here we suppose $l = 0, a \geqslant 0$, as in § 3. Now add the arc (t, s) to the network with $l(t, s) = 0, c(t, s)$ large, and $a(t, s)$ negatively large. If we start with the zero circulation and all node numbers zero, as in § 3, then the only out-of-kilter arc is (t, s) (it is in state γ_1) and hence it remains the only out-of-kilter arc throughout the computation. Then the origin for the labeling process is always s, the terminal t, and the labeling rules, flow change, and node number change all reduce to those of § 3.

References

1. R. Bellman, "On a Routing Problem," *Quart. Appl. Math.* 16 (1958), 87–90.
2. C. Berge, *Theorie des Graphes et ses Applications*, Dunod, Paris, 1958.
3. R. G. Busacker and P. J. Gowen, "A Procedure for Determining a Family of Minimal-Cost Network Flow Patterns," O.R.O. Technical Paper 15, 1961.
4. A. S. Cahn, "The Warehouse Problem," *Bull. Amer. Math. Soc.* 54 (1948), 1073 (abstract).
5. T. F. Cartaino and S. E. Dreyfus, "Application of Dynamic Programming to the Airplane Minimum Time-to-climb Problem," *Aero. Engr. Rev.* 16 (1957), 74–77.
6. A. Charnes and W. W. Cooper, "Generalizations of the Warehousing Model," *Op. Res. Quart.* 6 (1955), 131–172.

III. MINIMAL COST FLOW PROBLEMS

7. G. B. Dantzig, "Application of the Simplex Method to a Transportation Problem," *Activity Analysis of Production and Allocation*, Cowles Commission Monograph 13, Wiley, 1951, 359–373.

8. ———, "Upper Bounds, Secondary Constraints, and Block Triangularity in Linear Programming," *Econometrica* 23 (1955), 174–183.

9. ———, "Discrete Variable Extremum Problems," *Op. Res.* 5 (1957), 266–277.

10. ———, "On the Status of Multi-stage Linear Programming Problems," *I.S.I. Bull.* 36, 303–320.

11. ———, L. R. Ford, Jr., and D. R. Fulkerson, "A Primal-dual Algorithm," *Linear Inequalities and Related Systems*, Annals of Mathematics Study 38, Princeton University Press, 1956, 171–181.

12. G. B. Dantzig, D. R. Fulkerson, and S. Johnson, "Solution of a Large Scale Traveling Salesman Problem," *Op. Res.* 2 (1954), 393–410.

13. J. B. Dennis, *Mathematical Programming and Electrical Networks*, Wiley, New York, 1959.

14. S. E. Dreyfus, "An Analytic Solution of the Warehouse Problem," *Management Sci.* 4 (1957), 99–104.

15. A. L. Dulmage and I. Halperin, "On a Theorem of Frobenius-König and J. von Neumann's Game of Hide and Seek," *Trans. Royal Soc. Canada* 49 (1955), 23–29.

16. P. S. Dwyer, "The Solution of the Hitchcock Transportation Problem with a Method of Reduced Matrices," University of Michigan, 1955.

17. J. Egerváry, "Matrixok Kombinatorikus Tulajdonságairól," *Mat. es Fiz. Lapok* 38 (1931), 16–28. Translation by H. W. Kuhn, "On Combinatorial Properties of Matrices," George Washington University Logistic Papers 11 (1955).

18. M. M. Flood, "On the Hitchcock Distribution Problem," *Pacific J. Math.* 3 (1953), 369–386.

19. ———, "A Computational Algorithm for the Hitchcock Distribution Problem, University of Michigan (1959).

20. L. R. Ford, Jr., "Network Flow Theory," The RAND Corporation, Paper P-923, July 14, 1956.

21. ——— and D. R. Fulkerson, "A Simple Algorithm for Finding Maximal Network Flows and an Application to the Hitchcock Problem," *Canad. J. Math.* 9 (1957), 210–218.

22. ———, "Solving the Transportation Problem, *Management Sci.* 3 (1956), 24–32.

23. ———, "A Primal-dual Algorithm for the Capacitated Hitchcock Problem," *Naval Res. Logist. Quart.* 4 (1957), 47–54.

24. ———, "Constructing Maximal Dynamic Flows from Static Flows," *Op. Res.* 6 (1958), 419–433.

25. T. Fujisawa, "A Computational Method for the Transportation Problem on a Network," *J. Op. Res. Soc. Japan* 1 (1959), 157–173.

26. D. R. Fulkerson, "The Hitchcock Transportation Problem," The RAND Corporation, Paper P-890, July 9, 1956.

27. ———, "Increasing the Capacity of a Network: the Parametric Budget Problem," *Management Sci.* 5 (1959), 472–483.

28. ———, "An Out-of-Kilter Method for Minimal Cost Flow Problems," *J. Soc. Indust. Appl. Math.* 9 (1961), 18–27.

REFERENCES

29. D. R. Fulkerson, "A Network Flow Computation for Project Cost Curves," *Management Sci.* 7 (1961), 167–178.

30. ———, "On the Equivalence of the Capacity-constrained Transshipment Problem and the Hitchcock Problem," The RAND Corporation, Research Memorandum RM-2480, January 13, 1960.

31. J. W. Gaddum, A. J. Hoffman, and D. Sokolowsky, "On the Solution of the Caterer Problem," *Naval Res. Logist. Quart.* 1 (1954), 222–229.

32. D. Gale, "Transient Flows in Networks," *Michigan Math. J.* 6 (1959), 59–63.

33. ———, *The Theory of Linear Economic Models*, McGraw-Hill, 1960.

34. T. Gallai, "Maximum-minimum Sätze über Graphen," *Acta Math. Hung.* 9 (1958), 395–434.

35. ———, "Über Reguläre Kettengruppen," *Acta Math. Hung.* 10 (1959), 227–240.

36. B. A. Galler and P. S. Dwyer, "Translating the Method of Reduced Matrices to Machines," *Naval Res. Logist. Quart.* 4 (1957), 55–71.

37. A. N. Gleyzal, "An Algorithm for Solving the Transportation Problem," Research Paper 2583, *Nat. Bur. Standards J.* 54 (1955), 213–216.

38. I. Heller, "On Linear Systems with Integral Valued Solutions," George Washington University Logistics Seminar, 1956.

39. ———, "Constraint Matrices of Transportation-type Problems," *Naval Res. Logist. Quart.* 4 (1957), 73–78.

40. ———, "On the Traveling Salesman's Problem," *Proceedings Second Symposium in Linear Programming* 2 (1955), 643–665.

41. ——— and C. B. Tompkins, "An Extension of a Theorem of Dantzig," *Linear Inequalities and Related Systems*, Annals of Mathematics Study 38, Princeton University Press, 1956, 247–254.

42. F. L. Hitchcock, "The Distribution of a Product from Several Sources to Numerous Localities," *J. Math. Phys.* 20 (1941), 224–230.

43. A. J. Hoffman and J. B. Kruskal, Jr., "Integral Boundary Points of Convex Polyhedra," *Linear Inequalities and Related Systems*, Annals of Mathematics Study 38, Princeton University Press, 1956, 223–246.

44. M. Iri, "Algebraic and Topological Theory of Problems of Transportation Networks with the Help of Electric Circuit Models," *R.A.A.G. Research Note* 13, 1959.

45. ———, "A New Method of Solving Transportation-Network Problems," *J. Op. Res. Soc. Japan.* 3 (1960), 27–87.

46. W. Jacobs, "The Caterer Problem," *Naval Res. Logist. Quart.* 1 (1954), 154–165.

47. W. S. Jewell, "Warehousing and Distribution of a Seasonal Product," *Naval Res. Logist. Quart.* 4 (1957), 29–34.

48. ———, "Optimal Flow Through Networks," Interim Technical Report No. 8, Massachusetts Institute of Technology, 1958.

49. ———, "Optimal Flow Through Networks with Gains," presented at Second International Conference on Operations Research, Aix-en-Provence, France, 1960.

50. L. Kantorovitch, "On the Translocation of Masses," *Compt. Rend. (Doklady) Acad. Sci.* 37 (1942), 199–201.

51. ——— and M. K. Gavurin, "The Application of Mathematical Methods in Problems of Freight Flow Analysis," *Collection of Problems*

Concerned with Increasing the Effectiveness of Transports, Publication of the Akademii Nauk SSSR, Moskow-Leningrad, 1949, 110–138.

52. J. E. Kelley, Jr., "Critical Path Planning and Scheduling: Mathematical Basis," *Op. Res.* 9 (1961), 296-320.

53. ———— and M. R. Walker, "Critical Path Planning and Scheduling," Proc. of Eastern Joint Computer Conference, Boston, 1959.

54. T. C. Koopmans and S. Reiter, "A Model of Transportation," *Activity Analysis of Production and Allocation*, Cowles Commission Monograph 13, Wiley, 1951, 222–259.

55. D. König, *Theorie der Endlichen und Unendlichen Graphen*, Chelsea Publishing Co., New York, 1950.

56. H. W. Kuhn, "The Hungarian Method for the Assignment Problem," *Naval Res. Logist. Quart.* 2 (1955), 83–97.

57. ————, "Variants of the Hungarian Method for Assignment Problems," *Naval Res. Logist. Quart.* 3 (1956), 253–258.

58. G. J. Minty, "A Comment on the Shortest Route Problem," *Op. Res.* 5 (1957), 724.

59. ————, "Monotone Networks," *Proc. Roy. Soc. London*, Ser. A, 257 (1960), 194–212.

60. T. S. Motzkin, "The Assignment Problem," *Proceedings Sixth Symposium in Applied Mathematics*, McGraw-Hill, 1956, 109–125.

61. J. Munkres, "Algorithms for the Assignment and Transportation Problems," *J. Soc. Indust. Appl. Math.* 5 (1957), 32–38.

62. A. Orden, "The Transshipment Problem," *Management Sci.* 3 (1956), 276–285.

63. W. Prager, "On the Caterer Problem," *Management Sci.* 3 (1956), 15–23.

64. ————, "On Warehousing Problems," *Op. Res.* 5 (1957), 504–512.

65. L. W. Smith, Jr., "Current Status of the Industrial Use of Linear Programming," *Management Sci.* 2 (1956), 156–158.

66. A. N. Tolstoi, "Methods of Removing Irrational Shipments in Planning," *Sotsialisticheskii Transport* 9 (1939), 28–51.

67. A. W. Tucker, "Analogues of Kirchhoff's Laws," *George Washington University Logistic Paper* 3 (1950).

68. J. von Neumann, "A Certain Zero-sum Two-person Game Equivalent to the Optimal Assignment Problem," *Contributions to the Theory of Games*, Annals of Mathematics Study 28, Princeton University Press, 1953, 5–12.

69. D. F. Votaw, Jr., and A. Orden, "The Personnel Assignment Problem," Project SCOOP, Manual 10, 1952, 155–163.

70. H. M. Wagner, "On a Class of Capacitated Transportation Problems," *Management Sci.* 5 (1959), 304–318.

CHAPTER IV

MULTI-TERMINAL MAXIMAL FLOWS

Introduction

In this short concluding chapter we return to the topic discussed in Chapter I, but here a different point of view will predominate. Instead of focusing on the value of a maximal flow from one specified node to another, the primary concern will be with certain problems that arise when attention is shifted to all pairs of nodes. For example, how does one determine maximal flow values between all pairs of nodes in a network with capacity constraints on arcs? Does this necessitate solving all pairs of flow problems, or will something simpler suffice? Or, a more basic question: what are necessary and sufficient conditions for a given set of numbers to represent maximal flow values between pairs of nodes in some network? In addition to these questions, one other problem will be discussed: that of synthesizing a network which meets specified lower bounds on all maximal flow values, and at minimal total network capacity. These questions have been considered very recently by Mayeda [5], Chien [1], and Gomory and Hu [2]. Our exposition closely follows that of Gomory and Hu, who have given concise and elegant answers to all the questions posed above.

Throughout this chapter we shall deal only with undirected networks, for which the multi-terminal theory assumes a particularly simple and appealing form.

1. Forests, trees, and spanning subtrees

In this section we introduce and discuss briefly a few elementary notions concerning undirected graphs that have not been required heretofore. The first of these is that of a tree. A *tree* is simply a connected graph $G = [N; \mathscr{A}]$ that contains no cycles. Thus a tree has the property that there is a unique chain or path joining each pair of nodes, since the existence of two or more paths between the same pair of nodes implies the existence of a cycle in the graph. More generally, a graph, connected or not, without cycles, is called a *forest*; each connected piece of a forest is consequently a tree, when considered as a graph in its own right. It is easy to show, for example by induction on the number of nodes, that a tree

173

on n nodes has precisely $n - 1$ arcs. Indeed, any two of the three conditions:

(a) G is connected,
(b) G has no cycles,
(c) $|\mathscr{A}| = |N| - 1$,

implies the third and characterizes G as a tree.

Given a connected graph G on n nodes, one can delete arcs from G until a tree remains. Such a tree is called a *spanning subtree* of G. For example, a spanning subtree of the graph of Fig. 1.1 is shown in heavy arcs. If a graph G and a spanning subtree T of G are specified, we refer to the arcs of T as

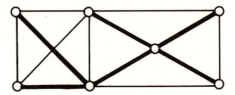

Figure 1.1

"in-tree" arcs, the others as "out-of-tree" arcs. Observe that if an out-of-tree arc is added to a spanning subtree, the resulting graph has just one cycle, consisting of the out-of-tree arc and the unique chain of in-tree arcs joining its end nodes. If any arc of this cycle is now deleted, the new graph is again a spanning subtree.

Suppose that each arc (x, y) of a connected graph G has associated with it a real number $a(x, y)$, which we might think of for the moment as the "length" of (x, y). Among all the spanning subtrees of G there is then a "longest" one; that is, one that maximizes the sum of the numbers $a(x, y)$ associated with arcs of the subtree (see Fig. 1.2 for an example). In studying multi-terminal network flows, maximal spanning subtrees turn out to be of considerable use. We shall therefore state and prove a maximality criterion for a spanning subtree, and then describe one of a number of simple algorithms that have been devised for constructing maximal spanning subtrees. We begin by noting an obvious necessary condition in order that a spanning subtree be maximal. Thus, suppose T is a maximal spanning subtree of G, and let $x_1, x_2, \ldots, x_k$ be the chain of in-T arcs joining x_1 and x_k. Here (x_1, x_k) is an out-of-T arc. Then clearly

$$(1.1) \qquad a(x_1, x_k) \leqslant \min [a(x_1, x_2), a(x_2, x_3), \ldots, a(x_{k-1}, x_k)],$$

for otherwise we could replace one of the in-T arcs of this chain by (x_1, x_k) to obtain a longer spanning subtree of G. On the other hand, if the condition (1.1) holds for each out-of-T arc, then the spanning subtree T is maximal.

This is not obvious, although it can be demonstrated in several ways. We shall sketch a proof showing, in fact, that if T_1 and T_2 are two spanning subtrees of G, and if each satisfies the assumption (1.1), then T_1 and T_2 are equal in length. This will certainly establish sufficiency. Given T_1 and T_2, we divide all of their arcs into three classes: arcs that belong to T_1 only (T_1-arcs), arcs that belong to T_2 only (T_2-arcs), and arcs that belong to both T_1 and T_2 (T_1, T_2-arcs). Suppose that T_1 and T_2 are distinct and take any T_2-arc, say (x_1, x_k), then look at the chain of in-T_1 arcs (x_1, x_2), $\ldots, (x_{k-1}, x_k)$; some of these, but not all, may be T_1, T_2-arcs. Thus there are T_1-arcs in this chain. By (1.1) applied to T_1, each of these T_1-arcs has length at least $a(x_1, x_k)$. We shall show that at least one of them has length equal to $a(x_1, x_k)$. For suppose each of them had length greater than $a(x_1, x_k)$. Then, taking each of them in turn, its end nodes are joined by a chain of in-T_2 arcs not containing the arc (x_1, x_k), since T_2 satisfies the hypothesis (1.1). It follows that T_2 contains a cycle, a contradiction. Hence some one of the T_1-arcs in the chain of in-T_1 arcs joining x_1 and x_k, say (x_p, x_{p+1}), has length equal to $a(x_1, x_k)$, as asserted. Now remove (x_p, x_{p+1}) from T_1 and replace it by (x_1, x_k). This yields a new spanning subtree T_1' that has the same length as T_1 and has one more arc in common with T_2. Moreover, the hypothesis (1.1) is again satisfied for T_1', as is readily verified. Hence the argument can be repeated, obtaining a succession of equal length trees $T_1, T_1', T_1'', \ldots$, the last of which is T_2. This proves

THEOREM 1.1. *A necessary and sufficient condition that a spanning subtree be maximal is that* (1.1) *hold for each out-of-tree arc.*

An analogous theorem holds for minimal spanning subtrees, as can be seen either directly or by replacing each $a(x, y)$ by its negative.

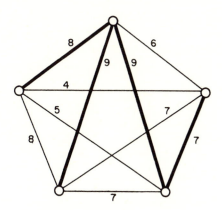

Figure 1.2

Kruskal [4] and Prim [6] have described several simple and direct algorithms for constructing a maximal (or minimal) spanning subtree of a given graph G. The validity of the following method, due to Kruskal, can be verified using Theorem 1.1. Begin by selecting a longest arc of G; at each successive stage, select (from all arcs not previously selected) a longest arc that completes no cycle with previously selected arcs, that is, keep the subgraph of selected arcs a forest at each stage. After $n - 1$ arcs have been selected, a longest spanning subtree has been constructed. For example, the construction might lead to the maximal spanning subtree indicated by heavy arcs in Fig. 1.2.

2. Realization conditions

For a given undirected network G with arc capacity function c, denote the maximal flow value from one node x to another node y by $v(x, y)$. Thus v is symmetric: $v(x, y) = v(y, x)$. We call v the *flow value function* of G, or more briefly, the *flow function*. (It is convenient in the sequel to put $v(x, x) = \infty$.) The first question that comes up is that of determining conditions under which a given symmetric function v can be realized as the flow function of some network. A first step in answering this question is provided by Lemma 2.1.

LEMMA 2.1. *If v is the flow function of a network, then for all nodes x, y, z,*

$$(2.1) \qquad\qquad v(x, y) \geqslant \min\, [v(x, z), v(z, y)].$$

Before proving Lemma 2.1, we note that condition (2.1), a kind of "triangle" inequality, puts severe limitations on the function v. For instance, applying (2.1) to each "side of the triangle" shows that, among the three functional values appearing in (2.1), two must be equal and the third no smaller than their common value. A further consequence of (2.1) is that if the network has n nodes, then v can have at most $n - 1$ numerically distinct functional values. We shall not prove this assertion here, since it will be a by-product of the proof of Lemma 2.2 below.

Notice that taking $v(x, x) = \infty$ eliminates the necessity of insisting that x, y, z be distinct in (2.1).

To prove Lemma 2.1, we use the max-flow min-cut theorem to pick out a minimal cut $(X, \overline{X})$ with x in X, y in $\overline{X}$, and $v(x, y) = c(X, \overline{X})$. Now z is either in X or $\overline{X}$. If z is in X, then

$$v(z, y) \leqslant c(X, \overline{X}) = v(x, y),$$

and (2.1) holds. If, on the other hand, z is in $\overline{X}$, then

$$v(x, z) \leqslant c(X, \overline{X}) = v(x, y),$$

and again (2.1) holds.

176

We point out that the proof uses the strong half of the max-flow min-cut theorem.

It follows inductively from (2.1) that

$$(2.2) \qquad v(x_1, x_k) \geqslant \min [v(x_1, x_2), v(x_2, x_3), \ldots, v(x_{k-1}, x_k)].$$

Here $x_1, x_2, \ldots, x_k$ is any sequence of nodes of the network.

The importance of conditions (2.1) is considerably enhanced by the fact that, not only are they necessary for realizability, they are also sufficient.

LEMMA 2.2. *If the non-negative, symmetric function v satisfies (2.1) for all* x, y, z, *there is an undirected network having flow function* v.

The discussion of § 1 can be brought into play in proving Lemma 2.2. Associate with each unordered pair (x, y) the number $v(x, y)$ to obtain an undirected graph each of whose arcs has a "length." Now let T be a maximal spanning subtree of this graph. It follows from (1.1) and (2.2) that if $x_1, x_2, \ldots, x_k$ is the chain of in-tree arcs from x_1 to x_k, then

$$(2.3) \qquad v(x_1, x_k) = \min [v(x_1, x_2), v(x_2, x_3), \ldots, v(x_{k-1}, x_k)].$$

Hence, if each in-tree arc is now assigned the capacity $c(x, y) = v(x, y)$, while each out-of-tree arc is deleted from the network, the flow network T has flow function v.

Thus if v is realizable, it is realizable by a tree.

We may summarize the discussion of this section in

THEOREM 2.3. *A non-negative symmetric function v is realizable as the flow function of an undirected network if and only if v satisfies (2.1). If v is realizable, it is realizable by a tree.*

3. Equivalent networks

We turn next to the problem of analysis of a flow network: to determine the flow function v in an efficient manner. We have just seen that v is realizable by a tree and hence that v can take on at most $n - 1$ numerically different values, where n is the number of nodes in the given network.

Suppose we call two n-node networks *flow-equivalent*, or briefly, *equivalent*, if they have the same flow function v. Thus every network is equivalent to a tree. Is there some way of constructing an equivalent tree that is better than first determining v explicitly by solving a large number of flow problems, and then constructing a v-maximal spanning tree?

Gomory and Hu have answered this question decidedly in the affirmative. Their procedure involves the successive solution of precisely $n - 1$ maximal flow problems. Moreover, many of these problems involve smaller networks than the original one. Thus one could hardly ask for anything better.

To begin the discussion of this method, let us suppose that a maximal flow problem has been solved with some node s as source, another node t as sink, thereby locating a minimal cut $(X, \overline{X})$ with s in X, t in $\overline{X}$ (see Fig. 3.1).

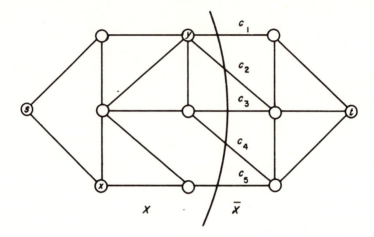

Figure 3.1

Suppose that we wish to find $v(x, y)$ where both x and y are on the same side of the s, t minimal cut $(X, \overline{X})$, say both x and y are in X. We first show that, for this purpose, all the nodes of $\overline{X}$ can be "condensed" into a single node to which all the arcs of the minimal cut are attached. (Several arcs joining the same pair of nodes can be replaced by a single arc, as in Fig. 3.2.) We call the network so obtained the *condensed* network. (Another

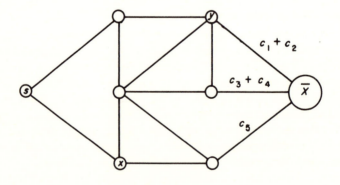

Figure 3.2

way of thinking of the condensed network is to imagine arcs joining all pairs of nodes of $\overline{X}$ with infinite capacity.)

178

LEMMA 3.1. *The maximal flow value $v'(x, y)$ between two ordinary nodes x and y of the condensed network is equal to the maximal flow value $v(x, y)$ in the original network.*

PROOF. Let $(Y, \bar{Y})$ be a minimal cut separating x and y in the original network and define sets

$$A = X \cap Y, \qquad \bar{A} = X \cap \bar{Y},$$
$$B = \bar{X} \cap Y, \qquad \bar{B} = \bar{X} \cap \bar{Y}.$$

Here $\bar{A}$ is the complement of A in X, $\bar{B}$ is the complement of B in $\bar{X}$. We may assume that $x \in A$, $y \in \bar{A}$, $s \in A$.

Case 1. $t \in B$. Now

$$c(X, \bar{X}) = c(A, B) + c(\bar{A}, B) + c(A, \bar{B}) + c(\bar{A}, \bar{B}),$$
$$c(Y, \bar{Y}) = c(A, \bar{A}) + c(A, \bar{B}) + c(B, \bar{A}) + c(B, \bar{B}).$$

Since $(Y, \bar{Y})$ is a minimal cut separating x and y, and since $(A \cup B \cup \bar{B}, \bar{A})$ separates x and y, we have

$$(3.1) \quad c(Y, \bar{Y}) - c(A \cup B \cup \bar{B}, \bar{A}) = c(A, \bar{B}) + c(B, \bar{B}) - c(\bar{A}, \bar{B}) \leqslant 0.$$

Since $(X, \bar{X})$ is a minimal cut separating s and t, and since $(A \cup \bar{A} \cup \bar{B}, B)$ separates s and t, then

$$(3.2) \quad c(X, \bar{X}) - c(A \cup \bar{A} \cup \bar{B}, B) = c(A, \bar{B}) + c(\bar{A}, \bar{B}) - c(\bar{B}, B) \leqslant 0.$$

Adding (3.1) and (3.2) shows that $c(A, \bar{B}) \leqslant 0$, and hence $c(A, \bar{B}) = 0$. It then follows from (3.1) and (3.2) that $c(B, \bar{B}) - c(\bar{A}, \bar{B}) = 0$ also. Hence $(A \cup B \cup \bar{B}, \bar{A}) = (A \cup X, \bar{A})$ is also a minimal cut separating x and y.

Case 2. $t \in \bar{B}$. A similar proof shows that $(A, \bar{A} \cup \bar{X})$ is a minimal cut separating x and y in this case.

In other words, there is always a minimal cut separating x and y such that the set of nodes $\bar{X}$ is on one side of this cut. Consequently, condensing $\bar{X}$ to a single node does not affect the value of a maximal flow from x to y.

Lemma 3.1 plays a fundamental role in the Gomory–Hu procedure for constructing an equivalent tree.

We now describe their construction.

Select two nodes arbitrarily and solve a maximal flow problem between them. This locates a minimal cut $(X, \bar{X})$, which we represent symbolically by two nodes connected by an arc of capacity $v_1 = c(X, \bar{X})$, as in Fig. 3.3.

Figure 3.3

179

In one node, the individual nodes of X are listed; in the other, those of $\bar{X}$. Next choose two nodes in X, say, and solve the resulting maximal flow problem in the $\bar{X}$-condensed network. The resulting minimal cut has capacity v_2, and is represented by an arc of this capacity connecting the two parts into which X is divided by the cut, say X_1 and X_2. The node $\bar{X}$ is attached to X_1 if it is in the same part of the cut as X_1; to X_2 otherwise. (See Fig. 3.4.)

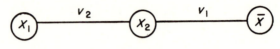

Figure 3.4

This process is continued. At each stage of the construction, some set Y, consisting of more than one node, is chosen from the tree diagram at that stage. The set Y will have a certain number of arcs attached to it in this tree. All of the sets (nodes of the tree) that can be reached from Y by paths using one of these arcs are condensed into a single node for the next maximal flow problem. This is done for each arc attached to Y in the tree. In the resulting network a maximal flow problem is solved between two nodes of Y. The set Y is partitioned into Y_1 and Y_2 by the minimal cut thus found; this is represented in the new tree by an arc having capacity equal to the cut capacity joining Y_1 and Y_2; the other nodes of the old tree are attached to Y_1 if they are in the Y_1 part of the cut; to Y_2 otherwise.

To illustrate the general step of the construction, suppose we have arrived at the tree diagram of Fig. 3.5, with Y to be split. Removal of the

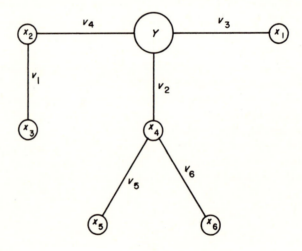

Figure 3.5

180

arcs attached to Y leaves the connected components Y; X_1; X_2, X_3; X_4, X_5, X_6. Then in the original network, the nodes of X_1 are condensed, as are those of $X_2 \cup X_3$, and $X_4 \cup X_5 \cup X_6$. Solving a maximal flow problem between two nodes of Y in the condensed network might then lead to the new tree shown in Fig. 3.6.

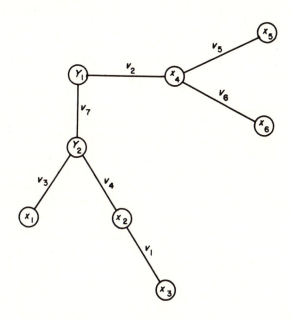

Figure 3.6

The process is repeated until all the sets consist of one node each. If the original network has n nodes, this point is reached after $n - 1$ maximal flow problems have been solved, since the final diagram is a tree on n nodes, each arc of which has been created by solving a flow problem. The number v_i attached to the ith arc of the final tree T is the capacity of this arc.

It is not at all evident that the tree T constructed in this manner is equivalent to the original network. That this is the case follows from Lemma 3.2.

LEMMA 3.2. *The maximal flow value between any two nodes of the original network is equal to*

$$\min (v_{i_1}, v_{i_2}, \ldots, v_{i_r}),$$

where $i_1, i_2, \ldots, i_r$ *are arcs of the unique path joining the two nodes in the final tree* T.

181

PROOF. Consider two nodes x and y. We show first that

$$(3.3) \qquad v(x, y) \leqslant \min (v_{i_1}, v_{i_2}, \ldots, v_{i_r}).$$

Here $i_1, i_2, \ldots, i_r$ are arcs of the path joining x and y in T. To see the validity of (3.3), it suffices to observe that the ith arc of T represents a cut $(X, \overline{X})$ in the original network having capacity v_i, and that the sets $X, \overline{X}$ are determined from T as follows. Delete the ith arc from T, leaving a forest of two trees; then X consists of all the nodes in one of these trees, $\overline{X}$ all the nodes of the other. This implies that x and y in (3.3) are separated by all the cuts corresponding to the arcs $i_1, i_2, \ldots, i_r$, and (3.3) follows. (That the final tree T does represent cuts in the manner described above is immediate from the construction, since each new tree produced in the construction represents cuts in this way provided the old tree does.)

To establish the reverse inequality is more difficult. This will be accomplished by showing that, at any stage of the construction, if an arc of capacity v joins nodes X and Y in the tree, then there is an x in X and a y in Y such that $v(x, y) = v$. This is certainly true at the first stage. We prove that the property is maintained. Consider a node Y about to be split, with X attached by an arc of capacity v. By the induction hypothesis there is an x in X and a y in Y with $v(x, y) = v$. Let s and t be the two nodes of Y for the next maximal flow problem. (We do not exclude the possibility $s = y$ or $t = y$ in what follows.) The set Y then divides into Y_1 and Y_2 with s in Y_1, t in Y_2. We may assume that X is attached to Y_1 (see Fig. 3.7).

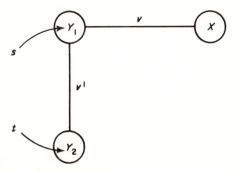

Figure 3.7

Of course s and t provide the two nodes such that $v(s, t) = v'$ for the new arc. As to the old arc of capacity v, there are two cases to consider. If y is in Y_1, then x and y provide the two nodes. The case in which y is in Y_2 is a little more troublesome. Here we shall show that s and x provide the required nodes. Notice that x and s are on one side of the minimal s, t cut of capacity v', and y and t are on the other. Thus, by Lemma 3.1, condensing

Y_2 to a single node in the original network does not change $v(x, s)$; that is, $v(x, s)$ remains unaffected if all pairs of nodes of Y_2 are joined by arcs of infinite capacity. Letting bars denote maximal flow values in the network thus obtained, we have

$$\bar{v}(x, s) = v(x, s),$$
$$\bar{v}(x, y) \geqslant v(x, y) = v,$$
$$\bar{v}(y, t) = \infty,$$
$$\bar{v}(t, s) \geqslant v(t, s) = v'.$$

Hence from (2.2),

$$\bar{v}(x, s) \geqslant \min\,[\bar{v}(x, y), \bar{v}(y, t), \bar{v}(t, s)] = \min\,[\bar{v}(x, y), \bar{v}(t, s)],$$

and consequently

$$v(x, s) = \bar{v}(x, s) \geqslant \min\,[v, v'].$$

Now $v' \geqslant v$ since the cut of capacity v' separates x and y. Hence $v(x, s) \geqslant v$. But equality must hold here because the cut of capacity v separates x and s. Thus $v(x, s) = v$, as was to be shown.

Consequently the capacities of the arcs in the final tree T actually represent maximal flow values between adjacent nodes of T. Using (2.2), this implies that

$$(3.4) \qquad v(x, y) \geqslant \min\,(v_{i_1}, v_{i_2}, \ldots, v_{i_r}),$$

where $i_1, i_2, \ldots, i_r$ are arcs of the path joining x and y in the final tree T. This, with (3.3), establishes Lemma 3.2 and shows that the construction produces an equivalent tree.

The beauty of the construction rests not only in the fact that an equivalent tree is produced with a minimum of effort, but also in the kind of equivalent tree; that is, one whose arcs represent the relevant $n - 1$ cuts in the original network. There are usually many trees equivalent to a network; for example, any maximal spanning subtree of the weighted graph corresponding to the flow function. In fact, it can be shown that every flow network is equivalent to a chain. But a tree produced by the construction is more than just equivalent to the starting network. Its structure corresponds precisely to the multi-terminal cut structure of the network. Gomory and Hu have called such a tree a *cut-tree* of the network. The same network may have more than one cut-tree, but in a sense this can happen only "by accident." That is, if we start with a connected graph G with arc capacity function c, and perturb its arc capacities by, say, adding ε^i to the ith arc capacity for small ε, then all cuts have distinct capacities, and the network will have a unique cut-tree. For surely only one set of cuts $(X_1, \bar{X}_1), (X_2, \bar{X}_2), \ldots, (X_{n-1}, \bar{X}_{n-1})$ corresponds to the $n - 1$ different values taken on by the flow function, and this list of cuts determines a unique cut-tree by the rule: nodes x and y are joined by an arc if and only if they lie on opposite sides of precisely one cut in the list. The arc (x, y) then has capacity equal to this cut capacity.

EXAMPLE.

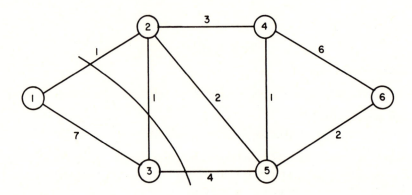

Figure 3.8

To begin the analysis for the network of Fig. 3.8, arbitrarily select nodes 1 and 6 for the first flow problem. This yields the cut ($\{1, 3\}, \{2, 4, 5, 6\}$) represented by the tree of Fig. 3.9.

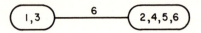

Figure 3.9

Taking 1 and 3 for the next flow problem and condensing 2, 4, 5, 6 gives the network of Fig. 3.10 and subsequent cut ($\{1\}, \{3, 2, 4, 5, 6\}$).

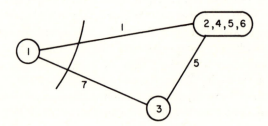

Figure 3.10

Hence the tree of Fig. 3.9 becomes Fig. 3.11.

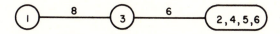

Figure 3.11

Next choose 2 and 4, giving the condensed network shown in Fig. 3.12 and cut ({1, 3, 2, 5}, {4, 6}).

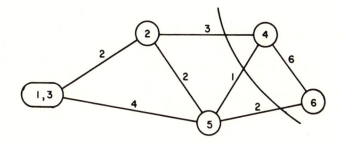

Figure 3.12

Hence Fig. 3.11 becomes Fig. 3.13.

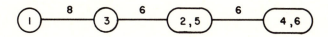

Figure 3.13

Selecting 2 and 5 for the next flow problem and condensing yields Fig. 3.14, with cut ({2}, {1, 3, 5, 4, 6}).

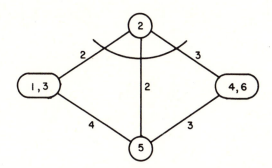

Figure 3.14

185

Thus the tree diagram at this stage is as shown in Fig. 3.15.

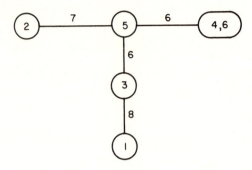

Figure 3.15

Finally choose 4 and 6 to get the condensed network of Fig. 3.16, and cut ({1, 2, 3, 5, 4}, {6}).

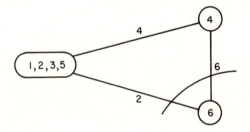

Figure 3.16

Consequently the final cut-tree is as shown in Fig. 3.17.

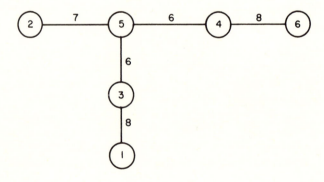

Figure 3.17

186

In the original network, the cuts picked out by the final cut-tree are shown in Fig. 3.18.

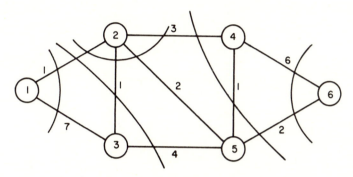

Figure 3.18

4. Network synthesis

Given a symmetric function r defined for all pairs of nodes of an n-node network, we shall call the network *feasible* if its flow function v satisfies

$$(4.1) \qquad v(x, y) \geqslant r(x, y), \qquad \text{all } x, y.$$

One problem that immediately suggests itself is that of constructing a feasible network which minimizes some prescribed function of the arc capacities, for example

$$(4.2) \qquad \sum_{x,y} a(x, y)c(x, y).$$

Here $a(x, y) = a(y, x)$ may be thought of as the known cost of installing one unit of arc capacity between x and y. This is a linear program, since the conditions (4.1) can be represented by writing down $2^{n-1} - 1$ linear inequalities:

$$(4.3) \qquad c(X, \overline{X}) \geqslant \max_{\substack{x \in X \\ y \in \overline{X}}} r(x, y),$$

corresponding to all cuts of the network. Of course, for even moderate values of n, the number of constraints makes it prohibitive to deal with this program explicitly. However, Gomory and Hu [3] have suggested simplex methods for the program that do not require an explicit enumeration and usage of all the constraints (4.3). We shall not discuss this rather general synthesis problem here, but shall look instead at the simpler version of the problem when all unit costs $a(x, y)$ are equal and may be assumed to be 1. For this problem, there is a remarkably simple and purely combinatorial method of synthesis.

187

To facilitate the description of this combinatorial method of synthesizing a minimal capacity feasible network, we shall carry along the example of Fig. 4.1, in which the requirements $r(x, y)$ are as indicated.

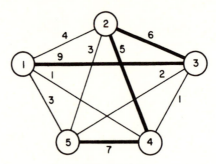

Figure 4.1

Let T be a *dominant requirement tree*; that is, a maximal spanning subtree of the requirement graph. We may construct T by the Kruskal algorithm, for instance. (In the example, a dominant requirement tree is shown by heavy arcs.) Then a necessary and sufficient condition that a network be feasible is that (4.1) hold for arcs of T. The necessity is of course obvious. For the sufficiency, suppose (x, z) is an out-of-T arc. Then it follows from (2.2), (4.1), and (1.1) that

(4.4)
$$v(x, z) \geqslant \min [v(x, y), v(y, u), \ldots, v(w, z)]$$
$$\geqslant \min [r(x, y), r(y, u), \ldots, r(w, z)]$$
$$\geqslant r(x, z).$$

Here $x, y, u, \ldots, w, z$ is the chain of in-T arcs joining x and z.

The synthesis uses only the dominant requirement tree T. First T is decomposed into a "sum" of a "uniform" requirement tree T plus a remainder (which is a forest of two or more trees) by subtracting the smallest in-T requirement from all other in-T requirements. Thus, the dominant requirement tree of Fig. 4.1 decomposes into that of Fig. 4.2.

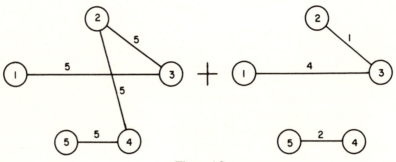

Figure 4.2

188

Each remaining non-uniform subtree is then further decomposed in the same way, and the process is repeated until T has been expressed as a sum of uniform requirement subtrees. In the example, this is achieved in one more step as shown in Fig. 4.3.

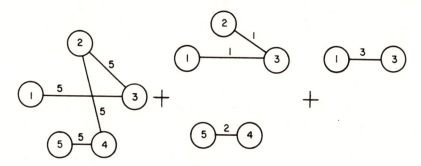

Figure 4.3

Each uniform tree of this decomposition is then synthesized by a cycle through its nodes (in any order), each arc of which has capacity equal to $\frac{1}{2}$ of the (uniform) requirement. (Clearly such a cycle will satisfy all requirements of a uniform tree.) The resulting cycles are then superposed to form a network G^*; that is, corresponding arc capacities are added. For example, doing this for Fig. 4.3 could give the cycles

1, 2, 3, 4, 5   (capacity $2\frac{1}{2}$)
1, 2, 3   (capacity $\frac{1}{2}$)
4, 5   (a single link of capacity 2)
1, 3   (a single link of capacity 3)

and resulting network G^* shown in Fig. 4.4.

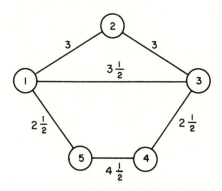

Figure 4.4

189

THEOREM 4.1.   *The network G^* is a feasible minimal capacity network.*

PROOF.   To see that G^* is feasible, it suffices to show that all requirements of the dominant tree T are met. This follows at once from the observations:

(a) a uniform requirement tree having requirements ε is synthesized by a cycle through its nodes, each arc of which has capacity $\varepsilon/2$,

(b) if two networks G' and G'' are superposed to form G, then $v(x, y) \geqslant v'(x, y) + v''(x, y)$.

It remains to prove that G^* is minimal; that is, the sum of the arc capacities of G^* is no larger than the corresponding sum for any feasible network. To this end, first define numbers, one for each node x,

$$(4.5) \qquad u(x) = \max_{y \neq x} r(x, y).$$

Thus $u(x)$ is the largest flow requirement out of x. (The same $u(x)$ results if the maximum in (4.5) is taken only over those y adjacent to x in the dominant requirement tree, as will be seen later on.) Now any feasible network with node set N and capacity function c must have

$$(4.6) \qquad c(x, N) \geqslant u(x);$$

hence for any feasible network,

$$(4.7) \qquad c(N, N) \geqslant u(N).$$

Here $c(N, N)$ is twice the sum of the arc capacities of the network. We shall show that the lower bound $u(N)$ in (4.7) is achieved by the network G^*; that is,

$$(4.8) \qquad c^*(N, N) = u(N),$$

whence G^* is a minimal capacity network.

To establish (4.8), first define $u'(x)$ as in (4.5), except that the maximum is taken over in-T arcs emanating from x. Then $u'(x) \leqslant u(x)$. But it is also clear from the construction of G^* that

$$(4.9) \qquad c^*(x, N) = u'(x),$$

and hence

$$(4.10) \qquad c^*(N, N) = u'(N) \leqslant u(N).$$

Hence $c^*(N, N) = u(N)$, as was to be shown. This completes the proof of Theorem 4.1.

It follows from this argument that $u'(x) = u(x)$, as was asserted earlier. This fact can also be seen directly without difficulty.

190

A consequence of Theorem 4.1 is: the linear program (4.2), (4.3), with all unit costs $a(x, y) = 1$, always has an optimal solution in which arc capacities are either integers or half-integers, provided the requirements are integers.

In general, there is a super-abundance of minimal capacity networks that can be obtained from the construction, since a uniform requirement tree can be synthesized by any cycle through its nodes. (Of course, if there are two distinct minimal capacity networks, there are infinitely many, because any convex combination of two such is also minimal.) Among all of these, there is one whose flow function dominates all others; that is, there is a feasible, minimal capacity network $\bar{G}$ such that

$$(4.11) \qquad\qquad \bar{v}(x, y) \geqslant v(x, y), \qquad\qquad \text{all } x, y,$$

where $v(x, y)$ is the flow function for any other feasible, minimal capacity network. To construct $\bar{G}$, one can go back to the original requirement network and revise each $r(x, y)$ upward to

$$(4.12) \qquad\qquad \bar{r}(x, y) = \min [u(x), u(y)].$$

Observe that this does not change the lower bound $u(N)$, since $\bar{u}(x) = u(x)$. If the requirements (4.12) are used in the synthesis, the network $\bar{G}$ thereby obtained meets all requirements exactly:

$$(4.13) \qquad\qquad \bar{v}(x, y) = \bar{r}(x, y).$$

For suppose $u(x) \leqslant u(y)$ and strict inequality held in (4.13). Then

$$\bar{c}(x, N) \geqslant \bar{v}(x, y) > u(x) = \bar{u}(x),$$

contradicting the fact that $\bar{c}(N, N)$ is equal to the lower bound $\bar{u}(N)$.

In the same way, one can see that $\bar{G}$ has the dominance property (4.11). For if there were a feasible minimal network G with

$$v(x, y) > \bar{v}(x, y) = \bar{r}(x, y)$$

and $\bar{r}(x, y) = u(x)$, say, then

$$c(x, N) \geqslant v(x, y) > u(x),$$

and the same contradiction results. In other words, more flow between any pair of nodes can be obtained only by increasing total network capacity.

References

1. R. T. Chien, "Synthesis of a Communication Net," *I.B.M. J.* 4 (1960), 311–320.
2. R. E. Gomory and T. C. Hu, "Multi-terminal Network Flows," I.B.M. Research Report, 1960; to appear in *J. Soc. Indust. Appl. Math.*

3. ———, "An Application of Generalized Linear Programming to Network Flows," I.B.M. Research Report, 1960.
4. J. B. Kruskal, Jr., "On the Shortest Spanning Subtree of a Graph and the Traveling Salesman Problem," *Proc. Amer. Math. Soc.* 7 (1956), 48–50.
5. W. Mayeda, "Terminal and Branch Capacity Matrices of a Communication Net," *I.R.E. Trans. on Circuit Theory* 7 (1960), 251–269.
6. R. C. Prim, "Shortest Connection Networks and Some Generalizations," *Bell System Tech. J.* 36 (1957), 1389–1401.

INDEX

OTHER RAND BOOKS

COLUMBIA UNIVERSITY PRESS, NEW YORK, NEW YORK

Bergson, Abram, and Hans Heymann, Jr., *Soviet National Income and Product, 1940-48*, 1954

Galenson, Walter, *Labor Productivity in Soviet and American Industry*, 1955

Hoeffding, Oleg, *Soviet National Income and Product in 1928*, 1954

THE FREE PRESS, GLENCOE, ILLINOIS

Dinerstein, Herbert S., and Leon Gouré, *Two Studies in Soviet Controls: Communism and the Russian Peasant; Moscow in Crisis*, 1955

Garthoff, Raymond L., *Soviet Military Doctrine*, 1953

Goldhamer, Herbert, and Andrew W. Marshall, *Psychosis and Civilization*, 1953

Leites, Nathan, *A Study of Bolshevism*, 1953

Leites, Nathan, and Elsa Bernaut, *Ritual of Liquidation: The Case of the Moscow Trials*, 1954

The RAND Corporation, *A Million Random Digits with 100,000 Normal Deviates*, 1955

HARVARD UNIVERSITY PRESS, CAMBRIDGE, MASSACHUSETTS

Bergson, Abram, *The Real National Income of Soviet Russia Since 1928*, 1961

Fainsod, Merle, *Smolensk under Soviet Rule*, 1958

Hitch, Charles J., and Roland McKean, *The Economics of Defense in the Nuclear Age*, 1960

Moorsteen, Richard, *Prices and Production of Machinery in the Soviet Union, 1928-1958*, 1962

THE MACMILLAN COMPANY, NEW YORK, NEW YORK

Dubyago, A. D., *The Determination of Orbits*, translated from the Russian by R. D. Burke, G. Gordon, L. N. Rowell, and F. T. Smith, 1961

O'Sullivan, J. J. (ed.), *Protective Construction in a Nuclear Age*, 1961

Whiting, Allen S., *China Crosses the Yalu: The Decision To Enter the Korean War*, 1960

MCGRAW-HILL BOOK COMPANY, INC., NEW YORK, NEW YORK

Bellman, Richard, *Introduction to Matrix Analysis*, 1960

Dorfman, Robert, Paul A. Samuelson, and Robert M. Solow, *Linear Programming and Economic Analysis*, 1958

Gale, David, *The Theory of Linear Economic Models*, 1960

Janis, Irving L., *Air War and Emotional Stress: Psychological Studies of Bombing and Civilian Defense*, 1951

Leites, Nathan, *The Operational Code of the Politburo*, 1951

McKinsey, J. C. C., *Introduction to the Theory of Games*, 1952

Mead, Margaret, *Soviet Attitudes toward Authority: An Interdisciplinary Approach to Problems of Soviet Character*, 1951

Scitovsky, Tibor, Edward Shaw, and Lorie Tarshis, *Mobilizing Resources for War: The Economic Alternatives*, 1951

Selznick, Philip, *The Organizational Weapon: A Study of Bolshevik Strategy and Tactics*, 1952

Shanley, F. R., *Weight-Strength Analysis of Aircraft Structures*, 1952

Williams, J. D., *The Compleat Strategyst: Being a Primer on the Theory of Games of Strategy*, 1954

THE MICROCARD FOUNDATION, MADISON, WISCONSIN

Baker, C. L., and F. J. Gruenberger, *The First Six Million Prime Numbers*, 1959

NORTH-HOLLAND PUBLISHING COMPANY, AMSTERDAM, HOLLAND

Arrow, Kenneth J., and Marvin Hoffenberg, *A Time Series Analysis of Interindustry Demands*, 1959

FREDERICK A. PRAEGER INC., NEW YORK, NEW YORK

Dinerstein, H. S., *War and the Soviet Union: Nuclear Weapons and the Revolution in Soviet Military and Political Thinking*, 1959

Speier, Hans, *Divided Berlin: The Anatomy of Soviet Political Blackmail*, 1961

Tanham, G. K., *Communist Revolutionary Warfare: The Viet Minh in Indochina*, 1961

PRENTICE-HALL, INC., ENGLEWOOD CLIFFS, NEW JERSEY

Dresher, Melvin, *Games of Strategy: Theory and Applications*, 1961

Hsieh, Alice L., *Communist China's Strategy in the Nuclear Era*, 1962

Newell, Allen (ed.), *Information Processing Language-V Manual*, 1961

PRINCETON UNIVERSITY PRESS, PRINCETON, NEW JERSEY

Baum, Warren C., *The French Economy and the State*, 1958

Bellman, Richard, *Adaptive Control Processes: A Guided Tour*, 1961

Bellman, Richard, *Dynamic Programming*, 1957

Bellman, Richard E., and Stuart E. Dreyfus, *Applied Dynamic Programming*, 1962

Brodie, Bernard, *Strategy in the Missile Age*, 1959

Davison, W. Phillips, *The Berlin Blockade: A Study in Cold War Politics*, 1958

Hastings, Cecil, Jr., *Approximations for Digital Computers*, 1955

Smith, Bruce Lannes, and Chitra M. Smith, *International Communication and Political Opinion: A Guide to the Literature*, 1956

Wolf, Charles, Jr., *Foreign Aid: Theory and Practice in Southern Asia*, 1960

PUBLIC AFFAIRS PRESS, WASHINGTON, D.C.

Krieger, F. J., *Behind the Sputniks: A Survey of Soviet Space Science*, 1958

Rush, Myron, *The Rise of Khrushchev*, 1958

RANDOM HOUSE, INC., NEW YORK, NEW YORK

Buchheim, Robert W., and the Staff of The RAND Corporation, *Space Handbook: Astronautics and Its Applications*, 1959

ROW, PETERSON AND COMPANY, EVANSTON, ILLINOIS

George, Alexander L., *Propaganda Analysis: A Study of Inferences Made from Nazi Propaganda in World War II*, 1959

Melnik, Constantin, and Nathan Leites, *The House without Windows: France Selects a President*, 1958

Speier, Hans, *German Rearmament and Atomic War: The Views of German Military and Political Leaders*, 1957

Speier, Hans, and W. Phillips Davison (eds.), *West German Leadership and Foreign Policy*, 1957

STANFORD UNIVERSITY PRESS, STANFORD, CALIFORNIA

Gouré, Leon, *The Siege of Leningrad*, 1962

Kecskemeti, Paul, *Strategic Surrender: The Politics of Victory and Defeat*, 1958

Kecskemeti, Paul, *The Unexpected Revolution: Social Forces in the Hungarian Uprising*, 1961

Kramish, Arnold, *Atomic Energy in the Soviet Union*, 1959

Leites, Nathan, *On the Game of Politics in France*, 1959

Trager, Frank N. (ed.), *Marxism in Southeast Asia: A Study of Four Countries*, 1959

UNIVERSITY OF CALIFORNIA PRESS, BERKELEY AND LOS ANGELES, CALIFORNIA

Gouré, Leon, *Civil Defense in the Soviet Union*, 1962

THE UNIVERSITY OF CHICAGO PRESS, CHICAGO, ILLINOIS

Hirshleifer, Jack, James C. DeHaven, and Jerome W. Milliman, *Water Supply: Economics, Technology, and Policy*, 1960

JOHN WILEY & SONS, INC., NEW YORK, NEW YORK

McKean, Roland N., *Efficiency in Government through Systems Analysis: With Emphasis on Water Resource Development*, 1958